AGRICULTURE ISSUES AND POLICIES

AGRICULTURAL PRODUCTION

Agriculture Issues and Policies

Additional books in this series can be found on Nova's website under the Series tab.

Additional E-books in this series can be found on Nova's website under the E-book tab.

AGRICULTURE ISSUES AND POLICIES

AGRICULTURAL PRODUCTION

FELIX C. WAGER

EDITOR

Nova Science Publishers, Inc.

New York

LIBRARY OF CONGRESS CATALOGING-IN-PUBLICATION DATA

Agricultural production / editor: Felix C. Wager.
p. cm.
Includes index.
ISBN 978-1-61668-695-6 (hardcover)
1. Agricultural productivity. 2. Agriculture--Economic aspects. I. Wager, Felix C.
S494.5.P75A37 2010
630--dc22

2010012150

Published by Nova Science Publishers, Inc. † New York

CONTENTS

PREFACE

Agriculture is the production of food and goods through farming and forestry and was the key development that led to the rise of human civilization, with the husbandry of domesticated animals and plants creating food surpluses that enabled the development of more densely populated and stratified societies. Agriculture encompasses a wide variety of specialties and techniques, including ways to expand the lands suitable for plant raising by digging water-channels and other forms of irrigation. This new book illustrates how the need for sustained agricultural production increases as the human population increases, while many natural resources grow scarce and the amount of land devoted to agriculture declines.

Chapter 1- The need for sustained agricultural production increases as the world's human population increases, many natural resources grow scarce, and the amount of land devoted to agriculture declines. For example, Vietnam loses 30,000 ha annually of prime rice land to urban development, yet it is the second highest exporter of rice in a world market that reached crisis levels during 2008 (Meerburg et al., 2009b). Between 1960 and 2000, the world's population doubled; in Asia alone the annual population growth until 2020 is estimated at 75 million, which is a lot of new mouths to feed (FAO, 2008). Hence, feeding the world's growing population continues to be a challenge for governments, especially in light of accelerated population growth, loss of agricultural land to urbanization and industrialization, shortage of agricultural labor due to migration of youth to cities, sustained economic growth leading to increase demands for meat protein (energy to produce 1 kg of meat protein requires 5 times that of proteins from cereals (Kawashima et al., 1997)), and pressures brought by climate change, loss of biodiversity, growing water scarcity, liberalized trade regimes, and inappropriate technology applications (e.g. growing of some food crops for bio-fuels). The future requires a sustainable agriculture base in which farms can produce food without causing severe or irreversible damage to ecosystem health.

Agro-ecosystems are complex systems that have transitioned from natural ecosystems by progressively incorporating interactions of three distinct systems: the ecological, the social/economic and the agricultural (Kogan and Lattin, 1999). Agro-ecosystems around the world range from modern, capital-intensive, large-scale monocultures to traditional, small, fragmented fields (e.g. in the Red River delta of northern Vietnam the average family holding is less than 0.25 ha, usually divided into 2 or 3 plots), and free-range livestock operations to confined livestock operations. Many other organisms compete with humans for food and natural resources, including wild animals, weeds, insects, and plant and animal pathogens. Although a wide array of vertebrate species cause damage in agriculture (Conover, 1998;

Olsen, 1998; Putman, 1989; Wywialowski, 1998), rodents pose one of the most serious threats to food production worldwide (Leirs, 2003; Meerburg et al., 2009b; Singleton et al., 2003; Stenseth et al., 2003). In this chapter, we review the rodent species involved, the types and levels of damage caused, the potential management options to reduce damage by rodents, and some research needs. Rodents also are carriers of >60 diseases that affect humans, some which can cause significant debilitation and can lead to affected smallholder farmers falling into an even greater poverty pit. We will not review rodent zoonoses but instead refer you to a recent review (Meerburg et al., 2009a).

Chapter 2- Transhumant pastoralism is an ecologically sound, a financially viable and a socially acceptable way of life for millions of the inhabitants of drylands. Strict observation of the carrying capacity of rangelands resulted in a sustainable management of the pastoralists' habitats for millennia. However, an unruly population growth and recurring prolonged droughts in recent years have devastated many mobile pastoralists and accelerated desertification of their domains. Provision of water, food, feed, fiber and fuelwood through spate irrigation along the pastoralists' trails may save many lives, decrease the number of environmental refugees, stem the city-ward migration, and sustain the transhumance tradition. Results of an 18-year study in a sandy desert, the previous wintering ground of nomads in southern Iran with a mean annual precipitation (MAP) of 243 mm, indicate that spate irrigation of a denuded rangeland increases its productivity both in wet (>200 mm MAP) and dry (<200 mm MAP) years. Mean usable forage yield was 490.77 and 182.53 kg ha^{-1} yr^{-1} for the spate-irrigated and control, respectively, in the wet years; this was 331.45 and 115.82 kg ha^{-1} yr^{-1} for the spate-irrigated and control, respectively, in the dry years; and 411.11(range: 136.27-721.50) and 149.18(range:78.00-350.20)kg ha^{-1} yr^{-1} for the spate-irrigated and control, respectively, during the 18-year period. Mean crown cover percentage for the mentioned treatments were 37.57 and 21.02 for the wet years; 26.97 and 17.46 for the dry years; and 32.27(range: 18.59-67.09) and 19.24 (range: 9.70-30.20) during the 18-year period. The higher yield ratio of spate-irrigated rangeland relative to control in droughts relative to wet years, 2.86 vs. 2.69, indicates the presence of a "pulse reserve" in the treated system. Moreover, contrary to the dictum of some ecologists, adapted annuals are recommended for rangeland improvement. Deposition of nutritious suspended load improved the available water holding capacity and fertility of sandy soils. Besides higher yield and better protection against wind and water erosion, the spate-irrigated land provides more nutritious feed and a longer grazing period. Furthermore, the artificial recharge of groundwater is achieved simultaneously if the spate-irrigated land is underlain with coarse-grained alluvium. Conversion of abandoned farm fields, whose productivity has been drastically impaired, to permanent grassland enhances the feed security of mobile graziers, while increases carbon sequestration, and thus mitigates the global warming.

Chapter 3- This paper introduces spreadsheet simulation models to evaluate the validity of national or regional disease surveys aimed at identifying infection in populations of farm animals. The process of evaluation includes specification or calculation of cluster-level diagnostic test sensitivity (the proportion of animals with the disease which test positive) and specificity (the proportion of animals without the disease which test negative), which are obtained from two probability distributions of the number of positive tests at individual-level expected from infected and non-infected clusters, respectively. Probability distributions for the number of positive clusters expected in a situation of the herd prevalence are specified and used to define survey properties (the survey being considered a diagnostic system), and receiver operating characteristic curves (consisting of a plot of sensitivity and specificity pairs for different cut-off values) are drawn. The result of a survey implemented to determine the prevalence of infectious bovine rhinotracheitis in dairy cattle in Uruguay, South America was used to illustrate this approach. The models can be adapted to a wide range of survey designs in animal health and production.

Chapter 4- Poultry production is a growing industry in Uruguay, South America. *Mycoplasma gallisepticum* (MG) and *Mycoplasma synoviae* (MS) are the etiological agent of chronic respiratory disease in chickens and may be horizontally and vertically infected. These economically important diseases can influence commercial poultry operations in Uruguay. Chickens infected with MG and/or MS become lessening in egg productivity, hatchability and feed efficiency and grow in mortality, waste carcass and veterinary costs. Increasingly, these possible performance and production deprivations require the strict control of MG and MS amongst all the poultry sectors. The objective of this study was to estimate the seroprevalence of the two pathogens mentioned above in relation to Mycoplasmosis in commercial chickens in Uruguay. Sera randomly selected from flocks of Uruguay were analysed using a commercial enzyme-linked immunosorbent assay (ELISA). Test prevalence were adjusted for the imperfect test characteristics using the Rogan–Gladen estimator (deterministic and stochastic functions) and Bayesian inference. Overall, 0.4% and 0.9% of the tested 1861 chickens were classified as test-positive for antibodies against MG and MS, respectively. Considering the simulated ELISA specificities of 99.5% for MG and 99.2% for MS and in the absence of other reported diagnoses from the area, these results were therefore considered to be "false-positives". The ELISA in combination with testing sera and statistical methods for true prevalence estimation provided a good survey tool at lower costs and time when compared to other approaches, and can be adapted to many other issues in animal health and production.

Chapter 5- The objective of this study was to estimate the true prevalence of seropositive individual sheep against *Chlamydophila abortus* and *Brucella ovis* using the Rogan–Gladen estimator in combination with Bayesian inference in Uruguay, South America. Ten ovine herds were kept under observation in 2008. Each study herd was randomly selected at different farms recruited from the Artigas (no. of the study herds = 1), Salto (no. of the study herds = 8) and Canelones (no. of the study herds = 1) Departments. The required total sample size of 1385 was determined by power analysis. Blood samples collected were used for diagnostic investigations. Sera were analysed using a commercial ELISA for the detection of antibody to *Chlamydophila abortus* and *Brucella ovis* in sheep serum. Adjusted outcomes were required for precise comparison of seroprevalence estimates. The overall antibody seroprevalence of *Chlamydophila abortus* and *Brucella ovis* would be extremely low. Considering the simulated ELISA specificities of 99.3% for *Chlamydophila abortus* and 97.0% for *Brucella ovis* and in the absence of other reported diagnoses from the area, these results were therefore considered to be "false–positives". The methodology was useful for obtaining estimates of *Chlamydophila abortus* and *Brucella ovis* prevalence and for establishing prevalence distributions which could be used as input parameters in risk assessment and decision models.

Chapter 6- The chapter evaluates the effects of land titling on the output and incomes of farmers in developing countries, especially focusing on the small farm sector. In response to the large numbers of small farmers who have lacked formal title and therefore secure tenure, a number of governments in developing countries have in recent years implemented comprehensive titling programs. The chapter will consider a range of such programs and the evidence of the benefits they have brought to the small farm sector in terms of investment, credit access and land markets. The chapter will as well examine factors which have restricted the impact of titling, including lack of capital access in the small farm sector, inadequate extension services, market outlets and insurance cover, and either limited or excessive access to credit. Also considered are issues and challenges involved in titling communal tribal lands, and how these challenges can be met. In summary, the chapter will contend that land titling makes an important contribution to increasing farm output, but its beneficial impact is contingent upon the existence of other factors that must be addressed in conjunction with the titling program itself.

Chapter 7- For the fulfillment of the thirsty ambition of self-sufficiency of the Soviets for cotton production, the arid Central Asian region, particularly, Uzbekistan has been extensively exploited. In fact, vast tracts of deserts have been converted into irrigated agricultural lands without proper consideration of the environment and technical standards. As a result, trends in natural resource degradation (soil salinity, desertification, water quality) as well as declining crop yields have dramatically increased.

The agricultural sector is the backbone for employment, food security and export revenues of the Central Asian countries. Since the independence of the Central Asian countries (after the breakup of the former Soviet Union), the situation has changed dramatically in terms of institutional, political and technical systems. Political transition, which is defined as a shift from a once planned centralized economy to a market-driven one, has introduced 'new' concepts like land tenure, water rights and different kinds of ownership. All such transformations have impacted the agricultural production in Central Asia.

The institutional change can be described as decentralization of the farming systems, i.e. transition from the former state collective farms into the smaller forms of private farms. The institutional interventions are aimed to increase agricultural production through improving water management. It is arguable that private production systems are the most effective business-driven forces but the situation is quite different in Central Asia due to the irrigated agriculture.

The biggest challenge for a sustainable irrigated agricultural production lies in the recent reforms of the water management sector in Central Asia and Uzbekistan. Water users' associations have been established to replace the former collective farming systems for irrigation water distribution and maintenance of irrigation infrastructures at the on-farm level. The intention of the national government was to shift the operation, maintenance and management of irrigation infrastructures to non governmental institutions (decentralization). However, these institutions have not fulfilled their promising tasks because of i) a rapid increase in the number of private farms along canals; ii) the cropping structure which is a mosaic with different crop water requirements compared to the former monoculture; iii) poor financial, training and technical capacities of the newly established institutions; iv) a state-ordered agricultural production quota system (for cotton and wheat).

This paper analyzes the historical aspects of the transformation in the farming production institutions in Central Asia with a special focus on Uzbekistan and comprehensively overviews the main current challenges facing the farming system and potential opportunities for reversing the situation.

Chapter 8- In the last two decades, the Chinese government has devoted large resources to insure that agricultural production keeps pace with a growing population and changing environment. This chapter focuses on seven areas influencing food production: 1) maintenance and expansion of the amount of arable land; 2) improvement and expansion of the irrigation system; 3) afforestation/reforestation, primarily to counter erosion and flooding; 4) creation of transgenic seeds of food crops that are pest- and drought-resistant; 5) other germ plasm improvements such as the development of hybrid varieties of rice and improvements of other varieties; 6) control of plant diseases and insect pests through cultural, chemical, and biological measures; 7) improvements of cultural (farming) practices, such as multiple and inter-cropping, use of wind tunnels and mulching. Analysis of these efforts suggests that insufficient attention has been paid to the areas most likely to significantly improve productivity, namely plant disease/pest control and application of improved cultural practices.

Chapter 9- The concept of Fair Trade is applied to the marketing of a variety of goods. In recent years it has met with a continually increasing interest of demanders. Different Fair Trade organisations are trying to accomplish an improvement of working and living conditions in countries where agricultural products stem from. They use Fair Trade certificates and pay a price evidently above world market standard. This trade concept is meant to permit basic social standards to be observed, especially in agricultural production. Fair Trade organisations make use of extrinsic product quality. Extrinsic product quality determines among other factors the consumer's perception of product quality. An important though thitherto poorly considered factor of extrinsic product quality is when the consumer cares about the production process, even if it does not affect product quality. The article deals with the questions how Fair Trade works and if the social aims in focus can be achieved by this trade concept. Our main result is that even though efficiency of redistribution through the Fair Trade institutions is lower than through traditional relief organisations the Fair Trade concept provides an additional incentive to pay for better living conditions in the third world. Fair Trade initiatives are unlikely to crowd out other redistribution channels. Moreover, there is a stimulus for producers to reorganize the production process in a more social manner even if this is not awarded by the fair trade company.

Chapter 10- This review concerns the possibility of using plant inhibitors of proteolytic enzymes to improve plant resistance to insects and phytopathogens. The main argument in favor of this approach is that protease inhibitors are widespread in plant tissues and highly active with respect to various proteases of insects, bacteria, and fungi. Results obtained for the last years point to perspective of genetic engineering approach. Main deficiencies and advantages of this approach and ways to improve its efficiency are discussed. Possibility of obtaining tobacco and potato plants resistant to bacterial infections by introduction of only one gene of serine proteinase inhibitor BWI-1a from buckwheat seeds through agrobacterial transformation of leaf disks has been demonstrated. It is noted that the discussed approach to increase plant tolerance to insects and pathogens has a number of advantages in comparison with traditional techniques and is ecologically pure.

In: Agricultural Production
Editors: Felix C. Wager, pp. 1-38

ISBN: 978-1-61668-695-6

Chapter 1

SUSTAINED AGRICULTURE: THE NEED TO MANAGE RODENT DAMAGE

Gary Witmer [*1] ***and Grant Single*** [*2]
[1]USDA National Wildlife Research Center, Fort Collins, Colorado, USA
[2]International Rice Research Institute, Metro Manila, Philippines

Introduction

The need for sustained agricultural production increases as the world's human population increases, many natural resources grow scarce, and the amount of land devoted to agriculture declines. For example, Vietnam loses 30,000 ha annually of prime rice land to urban development, yet it is the second highest exporter of rice in a world market that reached crisis levels during 2008 (Meerburg et al., 2009b). Between 1960 and 2000, the world's population doubled; in Asia alone the annual population growth until 2020 is estimated at 75 million, which is a lot of new mouths to feed (FAO, 2008). Hence, feeding the world's growing population continues to be a challenge for governments, especially in light of accelerated population growth, loss of agricultural land to urbanization and industrialization, shortage of agricultural labor due to migration of youth to cities, sustained economic growth leading to increase demands for meat protein (energy to produce 1 kg of meat protein requires 5 times that of proteins from cereals (Kawashima et al., 1997)), and pressures brought by climate change, loss of biodiversity, growing water scarcity, liberalized trade regimes, and inappropriate technology applications (e.g. growing of some food crops for bio-fuels). The future requires a sustainable agriculture base in which farms can produce food without causing severe or irreversible damage to ecosystem health.

Agro-ecosystems are complex systems that have transitioned from natural ecosystems by progressively incorporating interactions of three distinct systems: the ecological, the social/economic and the agricultural (Kogan and Lattin, 1999). Agro-ecosystems around the

---

* USDA National Wildlife Research Center, 4101 Laporte Avenue, Fort Collins, Colorado 80521 USA, gary.w.witmer@aphis.usda.gov
* International Rice Research Institute, Metro Manila, Philippines

world range from modern, capital-intensive, large-scale monocultures to traditional, small, fragmented fields (e.g. in the Red River delta of northern Vietnam the average family holding is less than 0.25 ha, usually divided into 2 or 3 plots), and free-range livestock operations to confined livestock operations. Many other organisms compete with humans for food and natural resources, including wild animals, weeds, insects, and plant and animal pathogens. Although a wide array of vertebrate species cause damage in agriculture (Conover, 1998; Olsen, 1998; Putman, 1989; Wywialowski, 1998), rodents pose one of the most serious threats to food production worldwide (Leirs, 2003; Meerburg et al., 2009b; Singleton et al., 2003; Stenseth et al., 2003). In this chapter, we review the rodent species involved, the types and levels of damage caused, the potential management options to reduce damage by rodents, and some research needs. Rodents also are carriers of >60 diseases that affect humans, some which can cause significant debilitation and can lead to affected smallholder farmers falling into an even greater poverty pit. We will not review rodent zoonoses but instead refer you to a recent review (Meerburg et al., 2009a).

THE NATURE OF RODENTS

Approximately 42% of all mammalian species in the world are rodents; this amounts to about 2,277 species of rodents (Wilson and Reeder, 2005). They occur on all continents with the possible exception of Antarctica. However, even there, commensal rodents may have been accidently introduced to the inhabited research stations. Rodent species have adapted to all life-styles: terrestrial, aquatic, arboreal, and fossorial (underground). Most rodent species are small, secretive, nocturnal, adaptable, and have keen senses of touch, taste, and smell. For most species of rodents, the incisors continually grow throughout their lifespan, requiring constant gnawing to keep the incisors sharp and at an appropriate length. In contrast to the normally small-sized body rodent, the capybara of South America can reach 70 kg in mass. Needless to say, a rodent this size can cause much damage to crops and rangeland (Ferraz et al., 2003; 2007). Alderton (1996) has written a fascinating account of the world of rodents and the love-hate relationship that has always existed and presumably always will between rodents and humans.

Rodents have ecological, scientific, social, and economic values (Dickman, 1999; Witmer et al., 1995). Rodents are important in seed and spore dispersal, pollination, seed predation, energy and nutrient cycling, the modification of plant succession and species composition, and as a food source for many predators. Additionally, some species provide food and fur for human uses, and can provide an ecosystem service for smallholder farmers through consuming pests of their crops (e.g. vermivores that feed on invertebrate pests in rice agro-ecosystems (Stuart et al., 2007)). Hence, the indiscriminate removal of rodents from ecosystems, including agro-ecosystems, is not the best management option in many cases (Aplin and Singleton, 2003; Brakes and Smith, 2005; Villa Cornejo et al., 1998).

Rodents are known for their high reproductive potential; however, there is much variability between species as to the age at first reproduction, size of litters, and the number of litters per year. In the topics and sub-tropics, reproduction can continue throughout the year, whereas, in more northerly latitudes, reproduction is usually seasonal and limited. Under favorable conditions, populations of some species such as the Microtines can irrupt, going from less than 10 per ha to a thousand per ha in the period of a few months (Korpimäki et al., 2004). During these periods of irruption, rodents will often invade crop fields and cause severe damage. From a management perspective, most rodent populations will exhibit a compensatory response to a severe population reduction with earlier age to sexual maturity, higher pregnancy rates, larger litter sizes, more litters per year, and a higher survival rate of young. Currently, there are no commercial products available to reduce the fertility of rodents although research in this area is on-going (discussed later in chapter).

As part of their life strategies, individuals of most rodent species have short life-spans and the annual mortality rate in a population is high, often about 70% (O'Brien, 1994; Singleton, 1989). Although rodents, generally, have good dispersal capabilities, unless conditions are very favorable, mortality rates during dispersal are quite high. Rodents succumb to starvation, predation, disease, drowning and other accidents, and various other mortality factors. Hence, most rodent species exhibit a classic r-selected life strategy: high reproductive rate with high mortality rate. An important management consideration is any quick reduction of a rodent population using lethal means (often with rodenticides as discussed later), will often result in a quick rebound of the population if no other actions are taken.

There are many interesting dynamics to various rodent populations that should be understood to better facilitate their management and to reduce damage (Batzli, 1992; Macdonald et al., 1999). Some populations go through annual cycles that may include high and low densities, active and inactive periods, reproductive and non-reproductive periods, and dispersal periods. To avoid inclement periods, some species exhibit winter dormancy (hibernation), while some species have summer dormancy (estivation) during hot, dry periods. Some species exhibit multi-year cycles; for example, the Microtines often reach population peaks (irruptions) every 3-5 years. While these cycles have been studied for decades, the driving factor(s) has not been definitively identified, but may involve long-term weather patterns, long-term nutrient cycles, predation, disease, and intra-specific social interactions (Krebs, 1996). During the early development of principles of population ecology, Charles Elton (1942), plus notable North American contemporaries such as Davis, Emlem and Howard (see Davis, 1987; Howard, 1988a), emphasized the importance of understanding the population biology of particular rodent species for effective management; one must take into consideration the specific demographics and capabilities of the species, along with the vagaries of cycles and periods of inactivity (e.g., Marsh, 1994) and the social and ecological context of modified agricultural landscapes (Singleton et al., 1999). Some recent reviews of the biology and ecology of pest rodents in the U.S. and control efforts include: pocket gophers (Marsh, 1992; Witmer and Engeman, 2007), ground squirrels (Marsh, 1994), voles (Witmer et al., 2009), and house mice (Witmer and Jojola, 2006).

RODENT DAMAGE AND THE SPECIES CAUSING DAMAGE

One of the serious threats to adequate world food production is the large volume of food production being consumed or contaminated by rodents. Some 280 million malnourished people could benefit if pre- and post-harvest losses by rodents are reduced (Meerburg et al.,

2009b). Fortunately, on a global-scale, only about 5-10% of the 2,277 species of rodents are serious agricultural pests (Witmer et al., 1995; Stenseth et al., 2003; Singleton et al., 2007a).

Table 1. Examples of rodents causing agricultural damage in various parts of the world.

Continent/Region	Genera	Crops Damaged	References
North America	*Cynomys, Geomys, Marmota, Microtus, Spermophilus, Thomomys*	Grains, alfalfa, flax, cotton, potato, sugar cane, trees	Marsh, 1984; 1988
Central & South America	*Ctenomys, Holochilus, Octodon, Oryzomys, Sigmodon, Zygodontomys*	Grains, beans, cotton, sugar cane, potato, cassava, bananas, trees	Elias and Fall, 1988; Hilje and Monge, 1988; Jackson, 1988
Europe	*Apodemus, Arvicola, Clethrionomys, Critetus, Microtus, Pitymys*	Grains, alfalfa, potato, beets, trees	Lund, 1984; 1988
Africa	*Arvicanthis, Mastomys, Meriones, Rhabdomys, Tatera, Thryonomys, Xerus*	Grains, cotton, sugar cane, potato, cassava, groundnut, plantation crops, trees	Fiedler, 1988; Smythe, 1986; Taylor, 1984
Middle East	*Hystrix, Meriones, Microtus, Nesokia, Psammomys, Spalax*	Grains, peanuts, root crops, vegetables, dates, trees	Moran and Keidar, 1993
South Asia	*Arvicola, Bandicota, Spermophilus, Cricetulus, Meriones, Microtus, Nesokia, Rattus, Tatera*	Grains, cotton, potato, sugar cane, vegetables, groundnut, banana, mango, pineapple, trees	Prakash, 1984; Prakash and Mathur, 1988; Zhi and Cheng-Xin, 1984
Southeast Asia/Pacific Islands	*Bandicota, Callosciurus, Mus, Rattus*	Grains, root crops, cocoa, sugar cane, oil palm, coconut, pineapple	Benigno and Sanchez, 1984 ; Hoque et al., 1988
Australia	*Rattus, Mus, Melomys*	Grains, sugar cane, macadamia nuts, banana, avocado, citrus	Caughley et al., 1998; Stenseth et al., 2003

On any given continent, there are generally about a half-dozen to a dozen genera that cause significant damage (Table 1). This list primarily includes native species of rodents, except for Australasia and some Asian countries. These species mainly cause damage to crops in the field. Additionally, there are several species of commensal rodents that cause damage mainly to physical structures (e.g., electrical wiring; fibre optics) and to food stuffs in storage by feeding and by contamination of stored food stuffs with their urine and feces (Ahmad et al., 1990; Proctor, 1994). The commensal rodents include the Norway rat (*Rattus norvegicus*), the ship or black rat (*R. rattus*), the Polynesian rat or kiore (*R. exulans*), and the house mouse (*Mus musculus* and *M. domesticus*). These species live in close proximity to humans, exploiting the favorable conditions that are created for them. As a result, they have spread throughout most of the world and cause significant losses of stored food stuffs. These rodents, along with some native rodent species, also pose health threats because of the many diseases they can carry that are transmissible to humans and livestock (Gratz, 1988; Meerburg et al., 2009a). Under some climatic conditions, commensal rodent populations will erupt, invade

Table 2. Levels of rodent damage to various crops around the world.

Region	Crop	Rodent	Damage Level	Reference(s)
North America	Wheat, alfalfa	Ground squirrel	25%	Askham, 1994
	Alfalfa	Vole	4%	O'Brien, 1994
	Peas	Vole	3-9%	Witmer et al., 2007c
	Corn	Rodents	1-100%	Hygnstrom et al., 1996
	Orchard fruit	Vole	35-66%	O'Brien, 1994
	Orchard fruit	Vole	36%	Askham, 1988
	Orchard trees	Vole	30%	Sullivan et al., 1987
	Rangeland forage	Kangaroo rat	11%	Hawthorne, 1994
	Rangeland forage	Prairie dog	18-90%	Hygnstrom and Virchow, 1994
	Rangeland forage	Pocket gopher	25%	Case and Jasch, 1994
	Stored grain contamination	House mice	76%	Timm, 1994a
	Stored grain	Norway rat	18%	Timm, 1994b
	Power outages	Tree squirrel	24%	Jackson, 1994
Central/South America	Beans	Rodents	5-10%	Meehan,1984
	Corn		26%	Ferraz et al., 2003
	Sugercane	Pocket gopher	21%	Villa-Cornejo, 2000
	Coconut	Rodents	77%	Elias and Fall, 1988
	Alfalfa	Rodents	10-30%	Elias and Fall, 1988
Europe	Sugarbeets	Wood mice	2-26%	Pelz, 1987
	Alfalfa	Vole	2-46%	Truszkowski, 1982
	Orchard trees	Vole	25%	Lund, 1988
	Horticulture	Voles	5-10%	Lund, 1988
	Stored grain contamination	Rodents	>90%	Meehan, 1984
	Farm fires	Rodents	50%	Richards, 1989
Africa	Corn	Rodents	20-30%	Ojwang and Oguge, 2003
	Corn	Rodents	26%	Bekele et al., 2003
	Sorghum	Multimammate rat	30%	Fiedler, 1988
	Rice	Rodents	80-100%	Fiedler, 1988
	Corn, sorghum, rice, legumes	Multimammate rat	48%	Mwanjabe et al., 2002
	Cacao	Rodents	4-12%	Fiedler, 1988
	Oil palm	Cane rat	23%	Fiedler, 1988
	Stored rice	Rodents	2-5%	Fiedler, 1988
	Rice	Rodents	7-30%	Prakash and Mathur, 1988
	Rice	Rodents	5-15%	Rao, 2003
	Wheat	Rodents	12-60%	Prakash and Mathur, 1988
	Sugarcane	Rodents	11%	Prakash and Mathur, 1988
	Corn	Rodents	1-14%	Prakash and Mathur, 1988
	Grains	Vole	50%	Wolf, 1977
	Groundnut	Rodents	4-26%	Prakash and Mathur, 1988
	Alfalfa	Vole	16-25%	Wolf, 1977
	Rangeland forage	Vole	80%	Nolte, 1996
Southeast Asia & Pacific Islands	Rice	Rodents	5-30%	Singleton, 2003
	Rice (Indonesia)	Rodents	16%	Singleton et al., 2005
	Rice	Rodents	5-27%	Hoque et al., 1988
	Corn	Rodents	5%	Hoque et al., 1988
	Sugarcane	Rodents	2-10%	Hoque et al., 1988
	Coconut	Rodents	12-65%	Hoque et al., 1988
	Pinapple	Rodents	10%	Hoque et al., 1988
	Cacao	Rodents	50-60%	Hoque et al., 1988
	Macadamia nut	Rodents	5-10%	Tobin, 1992
Australia	Macadamia nut	Rodents	30%	Caughley et al., 1998
	Sugarcane	Rodents	14-57%	Caughley et al., 1998
	Grains, sunflower	Rodents	12-25%	Brown et al., 2004

crop fields and pasture lands and cause significant damage. This has happens regularly with house mice in Australia (Brown et al., 2004; Caughley et al., 1998). There have been occasional irruptions of house mice in Hawaii (Tomich, 1986) and California (Pearson, 1963). There are irruptions of populations of rats and other native rodent species in India and Bangladesh (Chauhan, 2003), Laos (Khamphoukeo et al., 2003) and South America (Jaksic and Lima, 2003) associated with the flowering of bamboo. The episodic outbreaks in eastern India, Bangladesh and western Myanmar appear to be linked to a clonal species of bamboo that only blooms and sets fruit every 50+ years (Chauhan, 2003) and these outbreaks lead to significant food security issues (Belmain et al., 2008). Additionally, the commensal rodents have become established on many islands where they cause significant damage to natural resources and can lead to native species of animals, birds, and plants becoming endangered or extinct (Angel et al., 2009; Murphy et al., 1998; Witmer et al., 2007a; Witmer et al., 2007b).

The types of agricultural damage inflicted by rodents include the direct feeding on seeds and plants at all stages of the cropping cycle (e.i., planting, vegetative growth, maturation, and pre- and post-harvest). Additionally, rodents cause damage from their burrowing activities which can result in levee failures, flooding of fields, loss of water resources, and the undermining of structures and foundations (Joshi et al., 2000; Stuart et al., 2008). Burrows and burrow openings can result in damage to farm equipment and injury to workers or livestock. Through their gnawing activity, rodents can damage equipment, irrigation tubing, and buildings. For example, house mice cause significant damage to insulation in confined livestock operations (Hygnstrom, 1996). Chewing through wiring can result in power failure or devastating fires (Caughley et al., 1994). Rodents also compete with livestock for feed whether in confined operations or open rangeland. They also contaminate stored food with their feces and urine. Witmer and Engeman (2007) reviewed the many types of damage that can be caused by a single group of rodents, the fossorial pocket gophers.

Levels of Rodent Damage

In most agricultural settings, there is some level of rodent damage. Sometimes, the amount of damage may be small and considered inconsequential costs of business. Indeed, most farmers are not aware of damage to their growing cereal crops if the damage is less than 5%. However, in many situations, the damage is significant and the losses will threaten the peoples' livelihoods and food security (Belmain et al., 2008; UNDP, 2009), especially with a growing world population. In these cases, management actions are needed to reduce losses to tolerable levels. Examples of the levels of rodent damage to crops around the world are presented in Table 2. Rodent damage occurs in most parts of the world and many crops and resources are involved, and damage levels can be significant, if not severe. Damage is especially severe in tropical areas and in developing countries (Meehan, 1984; Singleton, 2003).

MONITORING RODENT POPULATIONS

An integrated pest management (IPM) approach generally will involve several methods woven into an effective damage reduction strategy (Witmer, 2007). An important principle of IPM is pest “scouting” (Matthews, 1996). However, the monitoring of vertebrate populations (especially small, nocturnal, secretive species) is problematic (Engeman and Witmer, 2000a). Monitoring first allows one to determine the specific species of rodents that occur in the area. Several to numerous rodent species may occur in any given area, but in many situations only one species is causing the damage. Knowing what species are present is important in designing a control strategy, to allow for the complications of baiting and trapping that other rodents may cause, and to plan for minimizing non-target losses. Monitoring rodent populations also is important because densities can fluctuate dramatically within a year and between years.

Obtaining accurate estimates of population density is difficult, as well as costly, in terms of labor, time, and resource requirements. There is considerable discussion within the wildlife profession as to the importance or need for highly accurate population density estimates in IPM programs because the objective is to manage damage rather than populations. Often, an index that efficiently tracks the rodent pest population is used. The index allows documenting of changes in the population through time and space, helps define the potential magnitude and geographical extent of damage that might result from population increases, and sets the stage for the implementation of an IPM strategy. Often, monitoring of pest populations is an important component of the assessment of the efficacy of control methods. There are a number of desirable properties to consider in the selection of a wildlife population indexing methods, including some associated with the planning stage, the in-field application, and the analytical phase (Witmer, 2005).

A wide array of methods exist for monitoring rodent populations and activity, including trap grids or transects, plot occupancy, open and closed holes per unit length for burrowing species, bait station or chew card activity, food removal, and runway or burrow opening counts (Engeman and Witmer, 2000a; Witmer and VerCauteren, 2001). Ideally the indices of choice have previously been validated as a reasonable measure of changes in population size for the species and habitat of interest. Such validations are available in some instances for rodents (e.g., Quy et al., 2009). Unfortunately, a good understanding of the relationship between the population index and the actual population density, or with the amount of crop/resource damage, is an exception rather than the rule (Leirs, 2003). There are advantages and disadvantages to each index that one must carefully consider before implementation. For example, the result of many indices can vary with the soil and habitat type (complexity, amount of cover, degree of human disturbance, etc), weather conditions, and the time of year. If the aim is to determine the efficacy of a management method then it is best to use 2 or more indices. Indeed, regulatory agencies may require two indices to be used, which is the case in the USA and the UK when data sets for rodenticide efficacy are submitted to federal regulatory agencies in support of new rodenticide registrations. Damage assessments are one of the most effective means of indexing population activity and determining program efficacy.

Challenges to Managing Rodent Damage and Populations

Rodents and their damage pose many management challenges. Solving rodent pest problems requires a careful consideration of:

1) the biology, population dynamics and seasonality of breeding of the pest species,
2) the ecology of the species within its physical and biotic environment,
3) an understanding of the relationships of the pest species to the activities of humans, including land uses, management practices, and other human activities,
4) the capacity (labor and financial resources) of farmers, government agencies, grain traders, etc., to implement and sustain the required management actions, and
5) the ecological consequences of the proposed actions (Singleton et al., 1999; Conover, 2002; Witmer, 2007).

It is only when we have adequate background knowledge in these areas that we can develop effective IPM strategies for rodent population and damage management (Figure 1). This is true because ecologically-based management requires us to focus on rodent population ecology, the environment effects for particular habitats, and the socio-economic factors that influence adoption (e.g., Witmer et al., 2003).

The traditional approaches to rodent population and damage management have relied on direct reduction of the population using rodenticide baits or rodent traps, and the reduction of habitat carrying capacity by habitat manipulation (Singleton et al., 2007a; Witmer et al., 1995). Today, many approaches focus on management efforts that are environmentally benign (Singleton et al., 1999; Pelz, 2003). Although many diverse techniques are available for rodent management (Table 3), most only provide temporary

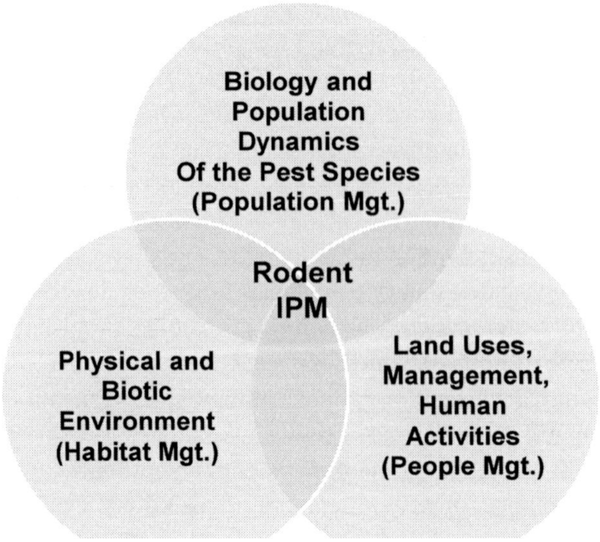

Figure 1. A triad of rodent population and damage management components that are underpinned by affordability and environmental effects.

control and/or are inhumane or adverse to the environment. Importantly, managers must consider the location, species, and type of damage before choosing an effective management strategy. Every situation can be unique and a "cookbook" approach will not suffice for every incident. Each method has
advantages and disadvantages, and generally using an IPM strategy will involve several methods to reduce damage. Many governments, universities, and non-governmental organizations have compiled books or manuals on rodent control specific to particular species and regions. Some examples from around the world are listed in Table 4.

Table 3. Methods and techniques for rodent control that have been suggested, tested, or used for various rodent problem situations (expanded from Fall, 1991).

Physical	Chemical	Biological	Other
Rodent proof construction	Baits/baiting systems	Fertility control	Appeasement
Passive barriers	Glues	Immunogens	Insurance
Electric barriers	Poison sprays	Habitat modification	Bounties
Drift fences	Poison moats	Cultural practices	Harvest
Trapping	Tracking powder	Crop timing	Compensation
Flooding burrows	Tracking greases, gel	Crop diversification, and species selection	
Drives	Repellents	Buffer crops	
Hunting	Attractants	Parasites	
Clubbing	Aversive agents	Diseases	
Frightening devices	Plant systematics	Predators	
Flame throwers	Sterilants	Ultrasonics	
Burrow destruction	Fumigation	Biosonics	
Habitat destruction	Psychotropic drugs	Resistant plants	
Harborage removal	Herbicides	Lethal genes	
Supplemental feeding	Poisons mixed with vehicle oil applied to flooded rice	Endophytic grasses	
Digging		Unpalatable plants	
Dogs together with flooding or digging			

Population Management of Rodents

Populations of rodents can be reduced by a variety of means. While methods such as trapping, burning, flooding, and drives have been---and are still being---used in developing countries, much of the world has come to rely on rodenticide baits for rodent control (Singleton et al., 1999; Witmer et al., 1995; Witmer and Eisemann, 2007). Most rodenticides were initially derived from naturally-occurring plant materials; however, most are now produced synthetically. Rodenticide delivery to targeted rodents typically occurs through consumption by rodents. There are two general classes of oral rodenticides. Acute rodenticides (e.g., compound 1080, zinc phosphide, strychnine) usually kill following a single feed. In contrast, chronic or multiple-feed rodenticides (e.g., warfarin, coumatetralyl, pindone, chlorophacinone, and diphacinone) usually require several days of feeding before an animal ingests a toxic dose. The distinction has become somewhat blurred because chronic rodenticides includes first (examples given) and second (e.g., bromadiolone, brodifacoum, difethialone) generation anticoagulants. Second generation compounds are very toxic and can usually kill following a single feeding, but still require several days for toxic symptoms to appear. Rodenticides can be applied in a variety of ways: hand-broadcast, aerially broadcast, placed in runways and burrows, placed near burrow openings, or placed in bait stations (Witmer and Eisemann, 2007). An additional group of rodent toxicants are the fumigants

Table 4. Examples of books and manuals providing rodent damage management guidelines.

Region	Title	Reference
Worldwide	Rodent Control in Agriculture	Greaves, 1982
Worldwide	Rats and Mice	Meehan, 1984
Worldwide	Control of Mammal Pests	Richards and Ku, 1987
Worldwide	Rodent Pest Management	Prakash, 1988
Worldwide	Rodent Pests and Their Control	Buckle and Smith, 1994
Worldwide	Ecologically-Based Rodent Management	Singleton et al., 1999
Worldwide	Rat, Mice and People: Rodent Biology and Management	Singleton et al., 2003
North America	Prevention and Control of Wildlife Damage	Hygnstrom et al., 1994
North America	Rodent Control	Corrigan, 2001
California USA	Vertebrate Pest Control handbook	Clark, 1994
Europe	Rodents as Pests	Putman, 1989
Africa	Rodent Pest Management in Eastern Africa	Fiedler, 1994
Africa	Rodent Biology and Integrated Pest Management in Africa	Leirs and Shockaert, 1997
Africa/Asia	Plant Protection Bulletin	FAO, 1988
Near East Asia	Rodent Pests and Their Control In the Near East	Greaves, 1989
Middle East Asia	Recent Advances in Rodent Control	Mohammad et al., 1983
India	Rodents in Indian Agriculture	Prakash and Ghosh, 1992
Bangladesh	Rodent Pests: Their Biology and Control in Bangladesh	Posamentier and Elsen, 1984
Pakistan	Handbook of Vertebrate Pest Control in Pakistan	Roberts, 1981
Pakistan	Vertebrate Pest Management	Brooks et al., 1990
Malaysia	Rodent Pests of Agricultural Crops in Malaysia	Khoo et al., 1982
South Asia	The Relative Importance of Crop Pests in South Asia	Geddes and Iles, 1991
Asia/Indonesia	Rodents and Rice	Quick, 1991
Asia/Philippines	Philippine Rats: Ecology and Management.	Singleton et al., 2008
Asia/Pacific	Field methods for rodent studies in Asia and the Indo-Pacific	Aplin et al., 2003
Australia	Managing Vertebrate Pests: Rodents	Caughley et al., 1998
Australia	MOUSER (Version 1.0): An interactive CD-ROM	Brown et al., 2003
New Zealand	The Ecology and Control of Rodents in new Zealand Nature Reserves	Dingwall et al., 1978
New Zealand	The Handbook of New Zealand Mammals (2nd edition) – see chapters on house mice, black rat, Pacific rat and Norway rat.	King, 2005

(e.g., gas cartridges, aluminum phosphide, methyl bromide) which are used in fumigating buildings or in burrow systems that are closed after application of the fumigant.

Considerable development has gone into making rodenticides effective, efficient, and relatively safe for use in buildings and surrounding areas. There also has been progress with the development of ecologically-based baiting strategies to assure safer and more effective use of rodenticides in cropland settings (Ahmed and Fiedler, 2002; Jackson, 2001; Ramsey and Wilson, 2000). However, primary and secondary poisoning is still a concern in croplands. In many countries, the use of rodenticides is carefully regulated by national, state, and provincial governments. Authorities decide who can use rodenticides and what training and record-keeping is required, along with which rodenticides and concentrations can be used and where, when, and how they are used. Research is underway to find new rodenticides as well as ways to make existing rodenticides more effective and less hazardous to non-targets and the environment. This is especially important in light of the fact that some rodenticides are being removed from the market and there are increasing restrictions on the use of many of these compounds (Jackson, 2001; Pelz, 2003; Witmer and Eisemann, 2007). Unfortunately, there are many cases of misuse of rodenticides or illegal or unapproved use (e.g., Schiller et al., 1999). In developing countries, there can also be issues of quality control in the production and sale of rodenticides (Bruggers et al., 1995).

There are many aspects of the biology and ecology of a rodent species that must be understood in order to effectively use rodenticides (or, for that matter, even traps or bait stations). Here we will present only a few important examples. Many rodent species are neophobic, exhibiting a fear of new objects, odors or tastes in their surroundings. As such, materials may have to be placed out for a few days to allow rodents to adjust to them. Traps may have to be placed in a locked-open position and baited for a few days before they are effective in catching rodents. This is also true with bait stations which may need to be in place for several days before rodents will enter them. Some traps are more effective in catching rodents than others and this varies widely by species of rodent. There has been a long history of rodent trap development; for example, Marsh (1997) reviewed the development and production of traps for pocket gophers. Some rodents become trap-shy after an initial capture and are difficult to re-capture, while others become "trap-happy" and can be readily re-captured. This becomes an important consideration for rodent researchers using capture-mark-recapture techniques to estimate population density.

Most rodents have well developed senses of taste and smell and relatively long memories. Consequently, baits must be fresh and not moldy or rancid. Additionally, some acute rodenticides are rather unpalatable (e.g., strychnine is bitter) and others (e.g., zinc phosphide) cause sickness so quickly that the animals may become bait-shy after an initial, non-lethal exposure. To avoid this, it is sometimes necessary to pre-bait with a non-toxic base material (e.g., grain without the zinc phosphide) before applying the toxic bait to help assure that the rodents will consume a lethal dose in a single feed. This is not a problem with anticoagulant rodenticides whereby the animal slowly becomes ill over time (i.e., as internal hemorrhaging begins), but continues to feed on the toxic bait which it does not associate with the gradual onset of illness. On the other hand, some populations of rodents that have been repeatedly exposed to an anticoagulant rodenticide, such as warfarin, have become resistant to the toxicant. Anticoagulant-resistant populations require the use of a different rodenticide or a different control strategy (e.g., Pelz, 2007). It is important to identify effective rodenticides for a particular species and situation, and an effective formulation and a baiting strategy that will effectively reduce the targeted population while minimizing non-target hazards (e.g., Cruz et al., 2008; Fan et al., 1999; Mathur, 1997; Moran, 2008; Witmer et al., 2007b).

It is also very important to recognize that rodent populations generally recover very quickly after rodenticide application (Zhong et al., 1999; Witmer et al., 2007c). Rodent populations can recover quickly even after major habitat alterations such as flooding (Zhang et al., 2007). Hence, continued applications or the use of other methods (discussed below) should be considered for the long-term control of rodent populations.

The feeding habits and food preferences of rodents may shift during the course of a year, therefore, baits used to deliver toxicants or to lure rodents into traps may have varied success depending on the seasonal preferences of the targeted rodents. For example, some rodents switch from a diet of green, succulent plant material early in the growing season to one primarily consisting of seeds once plants become senescent (Marsh, 1994).

The habitat needs, and especially cover requirements, for most rodents are critical because of the constant threat of predation, both day and night (see Ylönen et al., 2002). Knowing this, managers have tried to increase predator densities and reduce available cover as ways to reduce populations and damage. Unfortunately, prey populations usually drive predator populations, not the other way around. Artificial perches and nest boxes have been constructed to attract hawks and owls near croplands, orchards, and grasslands. Especially where natural perches were limited, these structures were used by raptors that preyed upon rodents and other animals (Ojwang and Oguge, 2003; Witmer et al., 2008a). In contrast, there is other evidence that suggests the rodent population or rodent damage is not substantially reduced as a result (e.g., Howard et al., 1985; Pelz, 2003; Sheffield et al., 2001).

Another theoretical way to reduce rodent populations is through disease agents or parasites. This approach has not yet had successes like those achieved during control for pest insect and plant populations. A major concern of using vertebrate biocides is that the agent may affect non-target species, including humans and livestock (Painter et al., 2004). This has been the case with the use of *Salmonella* spp. to control rats. A blood protozoan parasite, *Trypanosoma evansi* (Singla et al., 2003) and a liver nematode, *Capillaria hepatica (=Callodium hepaticum)* (Barker et al., 1991) have shown some potential for their ability to safely control rats and mice, however, the effect at the population level has not been sufficient to provide effective control. In Thailand, the protozoan, *Sarcocystis singaporensis*, is being investigated as a potential biocide (Boonsong, 1999; Khoprasert et al., 2008). A major problem is the maintenance of the disease agent or parasite in the environment after the target pest population has been greatly reduced. While there have been substantial successes with rabbit population control in Australia with the use of a myxoma virus and a rabbit calicivirus (Pech, 2000; Angulo and Bárcena, 2007), there are few success stories of biological control for mammal pest populations (see reviews by Leirs and Singleton, 2006; Baker et al., 2007).

Fertility control is often considered an attractive alternative to lethal control of rodents. There have been small-scale trials with various chemical compounds and some of these materials have shown promise (Miller et al., 1998). There are, however, many difficulties to overcome before any of these materials become available on the commercial market (Tyndale-Briscoe and Hinds, 2007; McLeod et al., 2007), including the need for an effective remote delivery system and the need to get a national, state, or /provincial registration that would allow the use of compounds in the field, especially given that the effects of such compounds would probably not be species-specific (Fagerstone et al., 2002). Using viruses as a vector for delivering species-specific sterility proteins has proven effective under laboratory conditions but the level of natural transmission to unaffected animals has been insufficient to proceed with field trials (Redwood et al., 2007). Nonetheless, several compounds and approaches have shown promise for fertility control of rodents (German, 1985; Seeley and Reynolds, 1989; Jacob et al., 2006; Zhao et al., 2007). There has also been some preliminary investigation of the ability of altered light cycles (e.g. artificial light at night in fields) to influence vole reproduction (Haim et al., 2004).

Ecological Relationships: Physical and Biotic Environment

Effective management of rodent pests also requires a thorough knowledge of the ecological relationships for the species in natural, semi-natural settings, and especially in human-altered settings (Leirs, 2003; Singleton et al., 2007b). For rodents, the physical environment is comprised of various structural features (e.g., soil, water bodies, rocks, plants, buildings, roads) and weather parameters. Densities of rodent populations will vary with regard to the environmental factors (e.g., soil type; Massawe et al., 2008). The biotic environment consists of all other species which can serve as competitors (e.g., other wildlife species, livestock, or humans) for food or space, or as predators (e.g., carnivores, raptors, snakes, humans). The biotic environment also includes endo-parasites, ecto-parasites, and disease organisms that can debilitate or kill rodents.

Some rodents can significantly alter their physical and biotic environment; for example, American beaver flood areas by building dams across streams or by plugging culverts, creating sizable water bodies and altering water flow regimes (Naiman et al., 1988). Also, pocket gophers can successfully prevent forest regeneration (after harvest, windstorm, or fire) by clipping and feeding on large numbers of tree seedlings (Engeman and Witmer, 2000b). On a smaller scale, rodents are very adept at creating burrow systems or sheltered nests (e.g., in trees, downed logs, or rock piles) to provide for their most basic cover needs. However, for the most part, rodents are at the vagaries of their physical and biotic environment (Batzli, 1992).

All rodents require food, shelter, and water. Availability and palatability of foods and quantity and quality of vegetative cover vary greatly between habitats and seasons, and sometimes between years (Tann et al., 1991). Consequently, rodents may switch their foraging preference and strategy during a year as well as between years. The success of many management activities directed towards rodents depend upon whether or not alternative foods are available. Additionally, rodents will often retreat to certain habitats or more sheltered areas when cover or food becomes sparse (e.g., after crop harvest; Singleton, 1989) or weather conditions become more severe. These areas serve as refugia and can act as source populations for population sustainability, increases, dispersal, or even irruptions (Elmouttie and Wilson, 2005; Giusti et al., 1996; Miño et al., 2007; Mills et al., 1991; White et al., 2003; Witmer et al., 2007c; Singleton et al., 2007b). Refugia shelter provides protection from predators and inclement weather as well as a favorable place to bear and rear their young. It has also been noted that taller vegetation generally supports higher rodent densities (Jacob, 2008; Sheffield et al., 2001; Witmer and Fantinato, 2003). Rodents also require water, but those requirements vary greatly by species. Some require no free-standing water at all and can meet their water needs through the metabolism of solid foods or the moisture on vegetation or other surfaces.

The amount and quality of food and vegetative cover are greatly influenced by precipitation, temperatures, photo-period, and other climatic parameters. There has been some progress in predicting and modeling rodent population responses to long-term weather patterns (e.g., house mouse irruptions in Australia [Pech el al., 1999], rodents in Africa [Leirs, 1999]). Generally, there are many factors involved and we have a relatively poor understanding of the interactions and rodent responses. We can rarely predict rodent density, where or when they will and if or when the populations will crash. This is why so many of our rodent management actions have been reactive rather than pro-active. Only with a better understanding of these underlying relationships will we begin to be more successful at predicting rodent populations and damage and be able to design and implement effective pro-active strategies (Leirs, 1999; Singleton et al., 1999; Stenseth et al., 2003).

Because of complex, and often poorly understood, ecological interactions between species, a focused attack on one rodent species may result in the unexpected. For example, Sullivan and others (1998) demonstrated this in vegetation management in orchards. They found that herbicide application to ground vegetation reduced vole numbers, but at the same time increased numbers of chipmunks and deer mice. Unpredicted outcomes can also result with efforts to alter or influence predator-prey relationships. In most situations, several rodent species usually occur and these may be in strong competition with each other. Hence, when one species is controlled or removed, another species which only occurred in low numbers may become much more numerous and begin to cause damage. This affect has been noted with control or eradication of introduced rats, whereby house mice populations suddenly irrupted once a competing species was removed (Corrigan, 2001; Witmer et al., 2007a). These undesired outcomes can be managed if proposed control programs take time beforehand to examine potential demographic dynamics at different trophic levels that involve invasive alien species and the invaded communities (Caut et al., 2009).

Influencing Food and Shelter to Reduce Rodent Populations or Damage

Because rodent food and cover (i.e., vegetation) can be greatly influenced by human activities, there has been considerable development of strategies to reduce populations and damage by manipulating vegetation. Many of these manipulations are not done just to reduce rodent habitat (which may be an incidental benefit) but for other reasons such as to reduce vegetative competition with crops or trees, to reduce soil pathogens, or to prepare sites for planting. Burning, plowing, disking, herbicide application all reduce vegetative cover, at least for the short term, and usually greatly reduce rodent populations (Massawe et al., 2003; Witmer, 2007). Plowing and disking have the additional advantage of disrupting the burrows of rodents. These methods have been used extensively in reforestation, orchards, and traditional agriculture. Understandably, farmers that have implemented no-till agricultural practices to reduce erosion, water loss and improve soil fertility have continued to suffer from high populations of rodents because the soil is not disturbed to an adequate depth and plant stubble (residues) are left on the surface (Witmer and VerCauteren, 2001; Witmer et al.,

2007c). Problems from rodents are compounded when grassy refugia are left along the periphery of crop fields that rodents can make use of when crop fields are rather bare (Brown et al., 2004). Additionally, a winter food supply for rodents is created by the spilled grains of crops such as wheat, barley, and legumes (Witmer et al., 2007c).

There has been some success in the use of lure crops or supplemental feeding to reduce damage by rodents or other vertebrates. Cracked corn or soybeans have been broadcast as lure crops after drill-seeding in no-till cropland to divert voles and other rodents from feeding on newly emerged crop seedlings or digging up and feeding on planted seeds (Witmer and VerCauteren, 2001). Sunflower seeds were broadcast on forest stands subject to tree squirrel damage with a subsequent reduction in tree damage (Sullivan and Klenner, 1993). A trap-barrier-system (TBS) was developed that uses some early planted crop fields to lure rodents into them (Singleton et al., 1998; several papers in Singleton et al., 1999; 2003). The lure fields are surrounded by a rodent barrier, but there are regularly spaced openings into multiple-capture rodent traps. The rodents in the traps are collected and killed daily. In some developing countries, the rodent carcasses are used as a source of high-protein food for humans and animals (Jacob et al., 2002; Jahn et al., 1999; Singleton et al., 2007a). This TBS method has reduced rodent invasion into the surrounding crop fields that are planted 2 to 3 weeks later. Aside from this clever use of multiple capture live traps, trapping for rodents is rarely effective or efficient in reducing populations over large acreages. One exception was coordinated community actions at a village level (100-200 ha) in intensively farmed rice fields in Southeast Asia where the average farm size was generally less than 1.5 ha (Singleton et al., 2005).

Another approach to vegetation manipulation still under investigation is the use of endophytic grasses. These are grass varieties that contain an alkaloid-producing fungus that can improve the hardiness of the grass and reduce herbivory. Some preliminary studies suggest that endophytic grass fields support lower rodent densities (Fortier et al., 2000; Pelton et al., 1991). These grasses could potentially be used in a variety of settings, but might be very valuable around cropfields and orchards where grassy areas have served as a traditional refugia for rodents and, hence, a source of dispersing individuals. Other species of unpalatable plants may offer a similar approach to lowering the rodent carrying capacity of a site (Giusti et al., 1996; Witmer and Fantinato, 2003).

Rodents compete for food with a variety of herbivores, including other rodent species, other wildlife, and livestock. There is some evidence that rodent populations can be reduced by intensive cattle or sheep grazing (Hunter, 1991; Moser and Witmer, 2000). In some cases, the intensive grazing can also reduce vegetative competition with tree saplings. In addition to reducing the food available to rodents, the livestock grazing may also compact the soil and disrupt burrow systems (Witmer and Fantinato, 2003).

Exclusion of Rodents from Areas or Resources

An alternative approach to reduce or eliminate rodent damage is to exclude them from high value areas. This is an attractive option in some situations because it is a nonlethal approach and could, potentially, solve the problem on a permanent basis. Exclusion devices can be physical barriers (e.g., fencing, sheet metal, or electric wires), frightening devices, ultrasonic or vibrating devices, or chemical repellents (Buckle and Smith, 1994; Hygnstrom et al., 1994). Unfortunately, it is very difficult to keep rodents out of an area that they want to enter. They can usually get over, around, under, or through any kind of barrier put in their way. Their small size, flexibility, agility, gnawing capability, along with their climbing and digging abilities make them a formidable adversary. They also habituate rather quickly to noxious odors, sounds, or lights (e.g., Timm, 2003). There are detailed guides available on how to rodent-proof buildings, but success is achieved only with much effort, expense, diligence, and maintenance (Corrigan, 2001; Baker et al., 1994). In open settings such as croplands or orchards, the task is much more difficult and the chance of success is much smaller. Although research in this area continues, there are few successes to report at this time (Pelz, 2003; Witmer et al., 2007c; Witmer et al., 2008b).

Short, low voltage, electric fences have been used with some success to exclude rodents from areas, but there were a number of concerns such as non-target hazards and excessive maintenance to keep the fences operating properly (Ahmed and Fiedler, 2002; Buckle and Smith, 1994; Shumake et al., 1979). Also, in Asia smallholder farmers cannot afford voltage regulators and instead some farmers directly run 220 volt power lines around their fields. This has led to deadly results not only for the rats but also for buffalo, goats and humans.

Physical barriers around individual tree seedlings have shown some success, but, again, there were concerns about cost, maintenance, and adverse effects on seedling growth (Marsh et al., 1990). Predator odors have shown some effectiveness in some trials for repelling rodents and other herbivores from areas or individual plants (Mason, 1998; Sullivan et al., 1988), but little effectiveness in other trials (e.g., Salatti et al., 1995). The sulfurous odors in predator urine, feces, glandular excretions, blood/bone meal, and putrescent eggs derived from the break-down of animal protein, all potentially serve as a cue to herbivores that a predator may be in the area and pose a threat to the herbivore (i,e., the potential prey; Mason, 1998). Another repellent that has shown some promise is capsaicin (a natural ingredient found in chili peppers), but a fairly high concentration ($\geq$ 2%) of this expensive material is usually needed for a reasonable level of effectiveness (Mason, 1998). These and other compounds have shown promise as rodent repellents (Ngowo et al., 2003; Oguge et al., 1997; Pelz, 2003; Witmer et al., 2001), but broad scale field use is still in its infancy.

Influence of Land Uses, Management Practices, and Human Activities on Rodent Populations

There are many things that landowners or managers can do to help reduce the risk of damage by rodents. An important first step is to familiarize themselves with the biology and ecology of the rodents (and other vertebrates that may cause damage) in the area, along with their signs of activity (burrow openings, mounds, runways, nests, tracks, droppings, gnawing patterns) and how to identify damage by those species (e.g., Hygnstrom et al., 1994). In North America, often information of this kind can be obtained at local or county extension offices or from other state/provincial or federal agencies. University wildlife damage specialists are also important sources of information. Unfortunately, in developing countries, wildlife damage management expertise is much less available (Singleton et al., 1999). A manual put together for identifying and working with rodents in Southeast Asia and the Pacific (Aplin et al., 2003) is an important step in overcoming those shortcomings. We provide a list of books and manuals that contain considerable region-specific information on damaging rodent species, the damage they cause, and management options (Table 4).

Proper sanitation around a property can significantly reduce food and cover available to rodents (Corrigan, 2001; Singleton et al., 1999). Removing rubbish piles, uncovered garbage receptacles, wood and metal debris piles, rock piles, piles or bales of hay, heavy mown grass, silage and other exposed livestock feed, grain spills, and mature tree fruit on the ground can aid in reducing rodent populations. A reduction in the availability of water (e.g., standing water or wet areas) can help, but is often difficult to achieve in an outdoor setting. Within buildings, food sanitation and removal of water sources are very important in the management of commensal rodents (Corrigan, 2001).

In some cases, agricultural producers have some discretion in the crops or crop varieties used, timing of planting, and the location and size of specific crop fields (Hannson, 1988; Singleton et al., 1999; Brown et al., 2004). Certain crops are more likely to be damaged than others. Cereal grains are more likely to be damaged by rodents than some crops such as soybeans or sunflowers (Brown et al., 2004; Mills et al., 1991; Wolf, 1977; Witmer and Fantinato, 2003). In many cases, large mono-culture crop fields will receive less rodent damage overall with most damage occurring along the periphery of the fields (Elmouttie and Wilson, 2005; Leirs, 2003; Mills et al., 1991; Witmer et al., 2007c); although in Asia the highest intensity of rodent damage is often in the center of fields, although the reason for this is unclear (Fall, 1977). Valuable crops that are especially vulnerable to rodents should not be grown near fallow areas, grasslands, or brushy areas that support rodent populations throughout the year and which serve as refugia from which rodents can rapidly disperse into crops.

In a region that is prone to periodic and substantial rodent damage, it is beneficial to have adjoining landowners cooperate in an overall strategy of reducing activities that support rodents and in rodent control activities (Jackson, 2001; Leirs et al., 1999; Posamentier, 1997; Singleton et al., 1999). Otherwise, a landowner may suffer continuous rodent damage despite rodent control efforts because the surrounding refugia in adjacent properties. Multiple landowner cooperation can help cost-sharing for rodent management activities and materials. In some situations, national, state, or local government support is available where vertebrate damage to agricultural production is severe.

Creation of a comprehensive and effective rodent damage management strategy would benefit from implementing an ecologically-based rodent management system (EBRM) that is tailored to the pest species, agricultural system, and prevailing climatic and habitat setting (Singleton et al., 1999). Several researchers and managers have developed this approach for use in developing countries. EBRM relies on a strong ecological understanding of the target pest species and the development of specific management actions at the farming systems level (Singleton, 1997; Singleton et al., 2007a). The key to EBRM is to reduce important resources needed by rodents such as food and nesting sites at critical times of the year through habitat modifications. Examples of these modifications would be synchrony of planting of crops (an important issue in Asia where holdings are small), minimizing the height and width of irrigation level banks to prevent rats from building nests, and controlling fallow vegetation along the edges of crop fields. Rodent populations are often controlled tactically at specific times of the year in specific habitats. This may be accomplished by various means (trapping, drives, bounties), but the emphasis is on a lower reliance on rodenticides and more community-wide approaches (Leirs, 2003; Singleton et al., 2007a). Versions of EBRM have been introduced into a number of countries to date:

- Thailand (Boonsong et al., 1999)
- China (Fan et al., 1999 ; Zhang et al., 1999 ; Zhong et al., 1999)
- Cambodia (Jahn et al., 1999)
- Lao PDR (Brown et al., 2007)
- Vietnam (Brown et al., 2006 ; Lan et al., 2003)
- Philippines (Cruz et al., 2003 ; Miller et al., 2008 ; Stuart et al., 2008)
- Indonesia (Leung et al., 1999 ; Singleton et al., 2005)
- Bangladesh (Belmain et al., 2006)
- Africa (Makundi et al., 1999)
- Australia (Brown et al., 2004)
- Europe (Pelz, 2003)

Perhaps the closest approach to EBRM in the USA was the effort of Engeman and Witmer (2000b) to predict and manage pocket gopher damage to reforestation. Similar ideas were pursued and recommended for vole control in agriculture and forestry in the USA (Witmer et al., 2009).

Examples of the kinds of management practices that farmers can implement to reduce losses to rodents were compiled by Brown and others (2004). The situation involves reoccurring but episodic house mice outbreaks in Australia with subsequent damage to various crops. Their list of recommendations included:

- Summer crop: early cultivation and weed control
- Winter crop: pre-sowing stubble management (burn), weed control
- Rice crop: stubble management (slash, graze, burn), bait stations, manage channels and banks
- Other actions: sow early, harvest cleanly, remove cover around sheds and silos, clean up grain spills, mouse-proof buildings, monitor mouse activity, bait key habitats using bait station.

Developing a rodent IPM (or EBRM) strategy requires the careful consideration of many factors (Andow and Rosset, 1990). Once the rodent species is correctly identified, it is important to monitor the status of the population and associated damage, using one or more of the many methods that exist. Is the rodent abundance related to the amount of damage that occurs and can a threshold be identified for when action should be initiated? Next, we should consider the nature of the rodent species and, its biology and ecology (particularly breeding ecology) in the setting in which the damage is occurring. How is the animal using its habitat? How is it interacting with other species? What are our actions doing to support the rodent population and to increase the amount of damage that occurs? What are our management options in terms of manipulating the rodent population, its habitat, and our activities and land use practices so that damage can be avoided or greatly reduced? What are the advantages and disadvantages of each of those management options? In general, it is best to start with the least invasive techniques before moving to more invasive techniques (e.g., Leirs, 2003; Pelz, 2003; Singleton et al., 2007a; Witmer, 2007). Finally, how do we mold all those considerations into a comprehensive rodent IPM and EBRM strategy that we can apply to the landscape?

The rodent management strategy under consideration should be evaluated for its ability to achieve the objective of rodent damage reduction within the set of real world constraints, including method effectiveness and duration, the associated cost and benefits, the legality, the socio-political acceptability, and whether the proposed actions are environmentally benign. Of course, once we apply the strategy, we should monitor the results to see if we have achieved the desired goal of damage avoidance or reduction (i.e., not just rodent population reduction), and whether or not there were unexpected results. The key here is to undertake adaptive management; the effectiveness of the management actions in the field are reviewed annually, if possible with the end users of the management, and changes made if required. Because relatively little is known about dealing with rodent damage situations in complex landscapes (e.g., agro-ecosystems, islands invaded by rodents), we are, in essence, conducting large-scale experimental field trials (Roy et al., 2009). It is only with adequate monitoring and adaptive resource management that we can interpret and learn from those trials and, ultimately develop a comprehensive and sustainable rodent IPM strategy.

Decision Support Systems

In some cases, once pest population or damage threshold levels are reached, decision support systems are in place to help the landowner or manager formulate and implement a pest damage control strategy (Coulson and Saunders, 1987). Unfortunately, there are relatively few such systems available and most are simple dichotomous keys and rudimentary computer programs. There is much variability in the goals, complexity, and input and output requirements and capabilities of existing rodent decision support systems. In particular, it is important to include economic considerations in animal crop damage, including benefit:cost analyses (Brown et al., 2004; 2006; Singleton et al., 2005; Sterner, 2008).

Important components of a comprehensive rodent decision support system include an overview of the species biology and ecology, population and damage identification and monitoring, damage potential and associated factors, an evaluation of alternative management techniques and the integration of techniques, a benefit-cost analysis component, computer user "friendliness" (for computer-assisted programs), and sources of additional information. We reviewed some available packages and noted their advantages and shortcomings (Table 5). A CD-ROM called Mouser (Brown et al., 2003), developed for house mouse irruptions in Australia, is the most complete rodent decision support systems that we have encountered, containing all of the desirable components. There is a great need, however, to improve most existing decision support systems and to develop many more for other rodent species, crops, fruit trees, etc., and to tailor them specifically for sets of end users.

Table 5. Evaluation of some small mammal decision support systems based on nine criteria.

Decision Key No.[1]	1	2	3	4	5	6	7	8	9	10
Criteria[2]										
Biology and Ecology	Yes	Yes	Yes	No	No	Yes	Yes	No	Yes	Yes
Monitoring	Yes	Yes	Yes	No	No	Yes	Yes	Yes	No	Yes
Damage Potential	Yes	Yes	Yes	No	Yes	Yes	Yes	No	Yes	Yes
Options	Yes	Yes	Yes	Yes	Yes	Yes	No	No	No	Yes
Benefit:Cost	No	Yes	No	Yes	No	Yes	Yes	No	Yes	No
Default Values	N/C[3]	Yes	Yes	Yes	Yes	Yes	Yes	Yes	Yes	No
Graphics	N/C	Yes	N/C	Yes	No	Yes	Yes	Yes	Yes	Yes
Save Option	N/C	Yes	N/C	No	No	Yes	Yes	Yes	Yes	N/C
Additional Information Sources	Yes	Yes	Yes	No	Yes	No	Yes	No	No	Yes

1 1=Timm, 1994a, house mice, buildings; 2=Brown et al., 2003, house mice, agricultural fields; 3=Engeman and Witmer, 2000b, pocket gophers, forestry; 4=Case and Timm, 1984, pocket gophers, alfalfa; 5=McComb, 1992, mountain beaver, forestry; 6=Salmon and Lickliter, 1983, ground squirrels, alfalfa; 7=Cox and Hygnstrom, 1993, prairie dogs, rangeland; 8=McGlinchy, 1999, possums, rabbits, all habitats; 9=Sterner, 2002, voles, alfalfa; 10=Aplin et al., 2003, rats in agricultural landscapes in Asia.

2 Important topics or modules and whether or not included in the package or publication.

3 N/C=Not Computer-assisted.

RESEARCH NEEDS AND FUTURE PROSPECTS

Effective rodent IPM strategies and decision support systems require substantial information that only long-term research of the given species and agro-ecosystem or commensal environment can provide. Furthermore, that research must be an integration of basic and applied studies with the needs of the end users and the desired impacts clearly defined before the research begins. Adequate information can result in more effective strategies, better predictive power, greater support and acceptance by the parties providing the funding, and by the end-users (e.g., farmers). Combining all this information is important to assure the application and sustainability of new strategies (Singleton et al., 1999). Unfortunately, there is relatively little support for long-term rodent research, and, in fact, there are relatively few rodent research scientists (Barnett, 1988). This situation is especially disturbing when one considers the imperative to manage food losses to rodents in developing countries. A recent report by the World Food Program "Emergency food security assessment" in northern Laos, reported that rodent outbreaks had a major impact on food security of smallholder farmers in 4 upland provinces with:

- 74% of interviewed households reporting losses to their crops of 50-100%

- 100% rice losses were common
- Major livelihood shock for rural households who rely on their own production for food

(Seehttp://home.wfp.org/stellent/groups/public/documents/ena/wfp202319.pdf)

While some new tools are being developed, many traditional tools for the control of vertebrate pests and their damage are being banned or restricted as the general public and legislators become increasingly active in land and resource management (Conover, 2002; Jackson, 2001). Examples include bans or restrictions on the use of rodenticide baits, traps, and field burning.

As suggested in the examples of this chapter, much more research is needed in both lethal and nonlethal means of resolving rodent damage in agricultural settings. For example, the prediction of rodent outbreaks so that proper measures can be taken to reduce the potential for damage would provide a valuable tool for rodent IPM (Leirs, 1999; Stenseth et al., 2003). While this is a difficult task without a detailed and time-honored data base, progress has been made in Africa (Leirs et al., 1996; Mwanjabe and Leirs, 1997), China (Zhang et al., 1999), and Australia (Pech et al., 1999; Krebs et al., 2004).

Efforts by researchers and research funding should be expanded to identify tools and strategies to reduce rodent populations and damage to agriculture (Howard, 1988b; Krebs, 1999; Witmer et al., 1995). Some areas of promising research directions include:

- Screen for varieties of crops that are less attractive to particular rodent species; or alternatively, identify varieties that are very attractive and use them to lure rodents to multiple capture traps
- Predicting rodent outbreaks/irruptions (as per house mice in Australia; Krebs et al., 2004)
- Protecting root systems from damage by tunneling rodents
- Effective rodenticides and methods to further reduce non-target animal hazards
- Unpalatable plants and endophytic grasses
- Effective and durable repellents
- Strategies for effective fertility control of rodents
- Species interactions with other native and non-native rodent species
- Food safety issues in agricultural areas, including better characterization of the losses caused by rodents to stored agricultural produce
- How integrated rodent management can influence the prevalence of rodent borne diseases that affect humans and their livestock
- Sociological studies of the factors that promote or hinder community actions that are required for effective broad scale ecologically based rodent management
- Anticipating changes to intensive cropping systems to meet increased food demands and developing management actions in accord with these changes
- Conducting active adaptive management to assist end users who themselves have changed management practices to cope with climate change
- More rigorous economic analysis of the costs and benefits of rodent IPM.

An additional concern, receiving more attention in recent years, regards who should pay for the cost of vertebrate pest population and damage management activities that benefit the general public or the agriculturalists of a region? Unfortunately, vertebrate damage, the cost of population and damage management, and management benefits are not evenly distributed across segments of the public and private sectors (Leirs et al., 1999; Posamentier, 1997). Additional research, increased public education, and increased sensitivity by public and private sector persons involved in vertebrate pest management may help resolve some of these problems.

Rodents, the damage they cause, and the diseases they transmit have plagued human populations since the beginning of civilization. There is no reason to believe that adverse interactions will not continue for the foreseeable future as these two groups vie for resources and co-evolve in natural and human-altered ecosystems, and especially in agro-ecosystems. Therein lies the challenge for practitioners of vertebrate IPM and EBRM.

References

Ahmad, E., Brooks, J. E., Munir, S., & Hussain, I. (1990). *Vertebrate pest management in grain storage*. Islamabad, Pakistan: Vertebrate Pest Control Project National Agricultural Research Centre. 48 pp.

Ahmed, M. S. & Fiedler, L. A. (2002). A comparison of four rodent control methods in Philippine experimental rice fields. *International Biodeterioration & Biodegradation* 49, 125-132.

Alderton, D. (1996). *Rodents of the World*. London, U. K.: Blandford Press. 192 pp.

Andow, D. A. & Rosset, P. M. (1990). Integrated Pest Management. In C. R. Carroll, J. H. Vandermeer, & P. Rosset (Eds.), *Agroecology*. (Pp. 413-439), New York, NY: McGraw – Hill Publishing Company.

Angel, A., Wanless, R. M., & Cooper, J. (2009). Review of impacts of the introduced house mouse on islands in the Southern Ocean: are mice equivalent to rats? *Biol. Invasions* 11, 1743-1754.

Angulo, E. & Bárcena, J. (2007). Towards a unique and transmissible vaccine against myxomatosis and rabbit haemorrhagic disease for rabbit populations. *Wildlife Research* 34, 567-577.

Aplin, K. P. & Singleton, G. R. (2003). Balancing rodent management and small mammal conservation in agricultural landscapes: challenges for the present and the future. In G. R. Singleton, L. A. Hinds, C. J. Krebs, & D. M. Spratt (Eds.), *Rats, mice and people: Rodent biology and management*. (Pp. 80-88), Canberra, Australia: Australian Centre for International Agricultural Research.

Aplin, K. P., Brown, P. R., Jacob, J., Krebs, C. J., & Singleton, G. R. (2003). *Field methods for rodent studies in Asia and the Indo-Pacific*. Canberra, Australia: ACIAR Monograph No. 100. 223 pp.

Askham, L. R. (1988). A two year study of the physical and economic impact of voles (*Microtus montanus*) on mixed maturity apple (*Malus spp.*) orchards in the Pacific Northwestern United States. In A. C. Crabb & R. E. Marsh (Eds.), *Proceedings of the*

13th Vertebrate Pest Conference (Pp. 151-155), Monterey, California: 13th Vertebrate Pest Conference (Mar 1-3, 1988).

Askham, L. R. (1994). Franklin, Richardson, Columbian, Washington, and Townsend Ground Squirrels. In S. E. Hygnstrom, R. M. Timm, & G. E. Larson (Eds.), *Prevention and Control of Wildlife Damage*. (Pp. B-159 – B-163), Lincoln, Nebraska: Cooperative Extension Division, University of Nebraska.

Baker, R. O., Bodman, G. R., & Timm, R. M. (1994). Rodent proof construction and exclusion methods. In S. E. Hygnstrom, R. M. Timm, & G. E. Larson (Eds.), *Prevention and Control of Wildlife Damage*. (Pp. B-137 – B-150), Lincoln, Nebraska: Cooperative Extension Division, University of Nebraska.

Baker, S. E., Singleton, G. R., & Smith, R. H. (2007). The Nature of the Beast: Using Biological Processes in Vertebrate Pest Management. In D. W. Macdonald & K. Service (Eds.), *Key Topics in Conservation Biology*. (Pp. 173-185), London, U.K.: Blackwell Publishing.

Barker, S. C., Singleton, G. R., & Spratt, D. M. (1991). Can the Nematode Capillaria Hepatica regulate abundance in wild house mice? Results of enclosure experiments in Southeastern Australia. *Parasitology* 103, 439-449.

Barnett, S. A. (1988). Ecology and economics of rodent pest management: The need for research. In I. Prakash (Ed.), *Rodent pest management*. (Pp. 459-464), Boca Raton, Florida: CRC Press, Inc.

Batzli, G. O. (1992). Dynamics of small mammal populations: a review. In D. McCullough & R. Barrett (Eds.), *Wildlife 2001: Populations*. (Pp. 831-850), New York: Elsevier Applied Science.

Belmain, S. R., Ali, M. S., Azad, A. K., Chakma, L. C., Dewan, U., Harun, M., Hasanuzzaman, A. T. M., Islam, S. M. F., Kadry, M. N. I., Kamal, N. Q., Sarker, N. J., Sarker, S. K., & Sarker, S. U. (2008). *Scientific assessment report on bamboo flowering, rodent outbreaks and food security: rodent ecology, pest management, and socio-economic impacts in the Chittagong Hill Tracts, Bangladesh*. Report of the United Nations Development Program, Bangladesh. 50 pp.

Belmain, S. R., Meyer, A., Kamal, N. Q., Shafali, R. B., Singleton, G., Aplin, K., & Mian, Y. (2006). Ecologically based rodent management for rice-based cropping systems in Bangladesh. In A. Sweetmore, F. Kimmins, & P. Silverside (Eds.), *Perspectives on pests II: Achievements of research under the UK Department for International Development Crop Protection Programme, 2000-05*. (Pp. 179-181), Aylesford, U.K.: Natural Resources International.

Bekele, A., Leirs, H., & Verhagen, R. (2003). Composition of rodents and damage estimates on maize farms at Ziway, Ethiopia. In G. R. Singleton, L. A. Hinds, C. J. Krebs, and D. M. Spratt (Eds.), *Rats, mice and people: Rodent biology and management*. (Pp. 262-263), Canberra, Australia: Australian Centre for International Agricultural Research.

Benigno, E. A. & Sanchez, F. F. (1984). Rodent problems in the Association of Southeast Asian nations. In A. C. Dubock (Ed.), *The Organization and Practice of Vertebrate Pest Control*. (Pp. 37-47), Surrey, U.K.: ICI.

Boonsong, P., Hongnark, S., Suasa-ard, K., Khoprasert, Y., Promkerd, P., Hamarit, G., Nookarn, P., & Jakel, T. (1999). Rodent management in Thailand. In G. R. Singleton, L. A. Hinds, H. Leirs, and Z. Zhang (Eds.), *Ecologically-based management of rodent pests*.

(Pp. 338-357), Canberra, Australia: Australian Centre for International Agricultural Research.

Brakes, C. R. & Smith, R. H. (2005). Exposure of non-target small mammals to rodenticides: short-term effects, recovery and implications for secondary poisoning. *Journal of Applied Ecology* 42, 118-128.

Brooks, J. E., Ahmad, E., Hussain, I., Munir, S., & Aziz Khan, A. (1990). Damage assessment at post-harvest level. In J. E. Brooks, E. Ahmad, I. Hussain, S. Munir, & A. Aziz Khan (Eds.), *Vertebrate pest management, a training manual.* (Pp. 55-58), Islamabad, Pakistan: Vertebrate Pest Control Project National Agricultural Research Centre Pakistan Agricultural Research Council.

Brown, P. R., Davies, M. J., Singleton, G. R., & Croft, J. D. (2004). Can farm-management practices reduce the impact of house mouse populations on crops in an irrigated farming system? *Wildlife Research* 31, 597-604.

Brown, P. R., Khamphoukeo, K., Sonekham, Invanh, Simone, Vison, Chanthi, Khamphamy, Lanfeuang, McWilliam, A., & Kingkeo, S. (2007). Development of rodent management strategies for the upland farming system in Lao PDR. *The Lao Journal of Agriculture and Forestry* July-December, 124-145.

Brown, P. R., Singleton, G. R., Norton, G. A., & Thompson, D. (2003). MOUSER (Version 1.0): An interactive CD-ROM for managing the impact of house mice in Australia. In G. R. Singleton, L. A. Hinds, C. J. Krebs, and D. M. Spratt (Eds.), *Rats, mice and people: Rodent biology and management.* (Pp. 264-266), Canberra, Australia: Australian Centre for International Agricultural Research.

Brown, P. R., Tuan, N. P., Singleton, G. R., Ha, P. T. T., Hoa, P. T., Hue, D. T., Tan, T. Q., Tuat, N. V., Jacob, J., & Muller, W. J. (2006). Ecologically based management of rodents in the real world: Applied to a mixed agroecosystem in Vietnam. *Ecological Applications* 16, 2000-2010.

Bruggers, R. L., Griffin, D. L., & Haque, M. E. (1995). Analysis of commercially available zinc phosphide from Bangladesh – implications for rodent control. *International Biodeterioration & Biodegradation* 36, 25-33.

Buckle, A. P. & Smith, R. H. (1994). *Rodent Pests and Their Control.* Wallingford, U.K.: CAB International. 405 pp.

Case, R. M. & Timm, R. M. (1984). Economic Model of Pocket Gopher Control. In W. F. Andelt (Ed.), *Proceedings of the Sixth Great Plains Wildlife Damage Control Workshop* (Pp. 53-56), Wichita, Kansas: Sixth Great Plains Wildlife Damage Control Workshop (December 12-14, 1983).

Case, R. M. & Jasch, B. A. (1994). Pocket Gophers. In S. E. Hygnstrom, R. M. Timm, & G. E. Larson (Eds.), *Prevention and Control of Wildlife Damage.* (Pp. B-17 – B-29), Lincoln, Nebraska: Cooperative Extension Division, University of Nebraska.

Caughley, J., Bomford, M., Parker, B., Sinclair, R., Griffiths, J., & Kelley, D. (1998). *Managing vertebrate pests: Rodents.* Canberra, Australia: Bureau of Resource Sciences and Grains Research and Development Corporation. 130 pp.

Caughley, J., Monamy, V. & Heiden, K. (1994). *Impact of the 1993 mouse plague.* Canberra, Australia: Grains Research & Development Corporation, Occasional Paper Series No. 7. 73 pp.

Caut, S., Angulo, E., & Courchamp, F. (2009). Avoiding surprise effects on Surprise Island: alien species control in a multitrophic level perspective. *Biological Invasions* 11, 1689–1703.

Chauhan, N. P. S. (2003). Observations of bamboo flowering and associated increases in rodent populations in the north-eastern region of India. In G. R. Singleton, L. A. Hinds, C. J. Krebs, and D. M. Spratt (Eds.), *Rats, mice and people: Rodent biology and management.* (Pp. 267-270), Canberra, Australia: Australian Centre for International Agricultural Research.

Clark, J. P. (1994). *Vertebrate Pest Control Handbook* (4th edition). Sacramento, California: Division of Plant Industry Integrated Pest Control Branch State of California Department of Food and Agriculture. 803 pp.

Conover, M. R. (1998). Perceptions of American agricultural producers about wildlife on their farms and ranches. *Wildlife Society Bulletin* 26, 597-604.

Conover, M. R. (2002). *Resolving Human-Wildlife Conflicts*. Boca Raton, Florida: CRC Press LLC. 418 pp.

Corrigan, R. M. (2001). *Rodent Control: A Practical Guide for Pest Management Professionals*. Cleveland, Ohio: GIE Media. 355 pp.

Coulson, R. N. & Saunders, M. C. (1987). Computer-assisted decision-making as applied to entomology. *Annual Review of Entomology* 32, 415-437.

Cox, M. & Hygnstrom, S. (1993). *Prairie Dog Control.* Lincoln, Nebraska: University of Nebraska. 15 pp.

Cruz, J., Leung, L. K-P., Lisle, A., Rivera, D. F., Staples, L., & Smith, M. (2008). Grain, pellet and wax block bait take by the house mouse (*Mus musculus*) and non-target species: Implications for mouse eradications on coral cay islands. *Integrative Zoology* 3, 227-234.

Cruz, M. S. D., Duque, U. G., Maquez, L. V., Martin, A. R., Joshi, R. C., & Tiongco, E. R. (2003). Rodent removal with trap-barrier Systems in a direct-seeded rice community in the Philippines. In G. R. Singleton, L. A. Hinds, C. J. Krebs, & D. M. Spratt (Eds.), *Rats, Mice and People: Rodent Biology and Management*. (Pp. 271-273), Canberra, Australia: ACIAR Monograph No. 96.

Davis D. E. (1987). Early behavioral research on populations. *American Zoologist* 27, 825–37.

Dickman, C. R. (1999). Rodent-ecosystem relationships: A review. In G. R. Singleton, L. A. Hinds, H. Leirs, & Z. Zhang (Eds.), *Ecologically-based management of rodent pests.* (Pp. 113-135), Canberra, Australia: Australian Centre for International Agricultural Research.

Dingwall, P. R., Atkinson, I. A. E., & Hay, C. (1978). The ecology and control of rodents in New Zealand nature reserves. In P. R. Dingwall, I. A. E. Atkinson, & C. Hay. *Proceedings of a Symposium convened by the Department of Lands and Survey*, (Pp. 1-237), Wellington, New Zealand: New Zealand Department of Lands and Survey (Nov 29-30, 1976).

Elias, D. J. & Fall, M. W. (1988). The rodent problem in Latin America. In I. Prakash (Ed.), *Rodent pest management*. (Pp. 13-28), Boca Raton, Florida: CRC Press, Inc.

Elmouttie, D. & Wilson, J. (2005). The potential importance of nut removal by rodents from Australian macadamia orchards. *Journal of Environmental Management* 77, 79-83.

Elton, C. (1942). *Voles, Mice and Lemmings: Problems in Population Dynamics.* London, U.K.: Oxford University Press. 496 pp.

Engeman, R. M. & Witmer, G. W. (2000a). IPM strategies: indexing difficult to monitor populations of pest species. In T. P. Salmon & A. C. Crabb (Eds.), *Proceedings of the 19th Vertebrate Pest Conference* (Pp. 183-189), San Diego, California: 19th Vertebrate Pest Conference (Mar 6-9, 2000).

Engeman, R. M. & Witmer, G. W. (2000b). Integrated management tactics for predicting and alleviating pocket gopher damage to conifer reforestation plantings. *Integrated Pest Management Reviews* 5, 41-55.

Fagerstone, K. A. (2002). Professional use of pesticides in wildlife management – An overview of professional wildlife damage management. In R. M. Timm & R. H. Schmidt (Eds.) *Proceedings of the 20th Vertebrate Pest Conference* (Pp. 253-260), Reno, Nevada: 20th Vertebrate Pest Conference (Mar 4-7, 2002).

Fall, M. W. (1977). Rodents in tropical rice. Tech. Bull. No. 36. College of Agriculture, University of the Philippines at Los Baños. 37 pp.

Fall, M. W. (1991). Controlling rice losses to rodents in rural communities. In G. R. Quick (Ed.), *Rodents and rice*. (Pp. 7-16), Los Banos, Philippines: Intl. Rice Research Institute.

Fan, N., Zhou, W., Wei, W., Wang, Q., & Jiang, Y. (1999). Rodent pest management in the Qinghai-Tibet alpine meadow ecosystem. In G. R. Singleton, L. A. Hinds, H. Leirs, & Z. Zhang (Eds.), *Ecologically-based management of rodent pests.* (Pp. 285-304), Canberra, Australia: Australian Centre for International Agricultural Research.

Ferraz, K. M. P. M. B., Ferraz, S. F. B., Moreira, J. R., Couto, H. T. Z., & Verdade, L. M. (2007). Capybara (*Hydrochoerus hydrochaeris*) distribution in agroecosystems: A cross-scale habitat analysis. *Journal of Biogeography* 34, 223-230.

Ferraz, K. M. P. M. B., Lechevalier, M. A., Couto, H. T. Z., & Verdade, L. M. (2003). Damage caused by capybara on a corn field. *Scientia Agricola* 60, 191-194.

Fiedler, L. A. (1994). Rodent pest management in eastern Africa. *FAO Plant Production and Protection Paper* 123. Rome, Italy.

Fiedler, L. A. (1988). Rodent problems in Africa. In I. Prakash (Ed.), *Rodent pest management*. (Pp. 35-65), Boca Raton, Florida: CRC Press, Inc.

Food and Agriculture Organization (FAO). (2008). *Feeding the world sustainable Management of Natural Resources*. Rome, Italy: Food and Agriculture Organization, United Nations. 19 pp.

Food and Agriculture Organization (FAO). (1988). *Plant Protection Bulletin* 36, 106-162.

Fortier, G., Bard, N., Jansen, M., & Clay, K. (2000). Effects of tall fescue endophyte infection and population density on growth and reproduction in prairie voles. *Journal of Wildlife Management* 64, 122-128.

Geddes, A. M. W. & Iles, M. (1991). The relative importance of crop pests in South Asia. *Natural Resources Institute Bulletin* No. 39. 102 pp.

German, A. (1985). Contact effect of diethylstilbestrol (DES) on the suppression of reproduction in the Levant vole, (*Microtus guentheri*). *Acta Zooligical Fennica* 173, 179-180.

Giusti, G. A., Whisson, D. A., & Gorenzel, W. P. (1996). Rodents and cover crops – a review. In R. M. Timm & A. C. Crabb (Eds.), *Proceedings of the 17th Vertebrate Pest Conference* (Pp. 59-61), Rohnert Park, California: 17th Vertebrate Pest Conference (Mar 5-7, 1996).

Gratz, N. G. (1988). Rodents and human disease: A global appreciation. In I. Prakash (Ed.), *Rodent Pest Management*. (Pp. 101-169), Boca Raton, Florida: CRC Press, Inc.

Greaves, J. H. (1982). Rodent control in agriculture. *FAO Plant Production and Protection Paper* 40. Rome, Italy.

Greaves, J. H. (1989). *Rodent pests and their control in the near east.* Rome, Italy: Food and Agriculture Organization of the United Nations. 112 pp.

Haim, A., Shanas, U., Zisapel, N., & Gilboa, A. (2004). Rodent pest control: Use of photoperiod manipulations as a tool. In C. J. Feare & D. P. Cowan (Eds.), *Advances in Vertebrate Pest Management III.* (Pp. 29-38), Furth, Germany: Filander Verlag.

Hansson, L. (1988). Natural resistance of plants to pest rodents. In I. Prakash (Ed.), *Rodent pest management.* (Pp. 391-397), Boca Raton, Florida: CRC Press, Inc.

Hawthorne, D. W. (1994). Cotton Rats. In S. E. Hygnstrom, R. M. Timm, & G. E. Larson (Eds.), *Prevention and Control of Wildlife Damage.* (Pp. B-97 – B-99), Lincoln, Nebraska: Cooperative Extension Division, University of Nebraska.

Hilje, L. & Monge, J. (1988). *Diagnostico Preliminar Acerca de los Animales Vertebrados que son Plagas en Costa Rica.* Heredia, Costa Rica: Universidad Nacional. 17 pp.

Hoque, M. M., Sanchez, F. F., & Benigno, E. A. (1988). Rodent problems in selected countries in southeast Asia and islands in the Pacific. In I. Prakash (Ed.), *Rodent pest management.* (Pp. 85-99), Boca Raton, Florida: CRC Press, Inc.

Howard, W. E. (1988a). Rodent pest management: The principles. In I. Prakash (Ed.), *Rodent pest management.* (Pp. 285-293), Boca Raton, Florida: CRC Press, Inc.

Howard, W. E. (1988b). Areas of further research. In I. Prakash (Ed.), *Rodent pest management.* (Pp. 451-458), Boca Raton, Florida: CRC Press, Inc.

Howard, W. E., Marsh, R. E., & Corbett, C. W. (1985). Raptor perches: their influence on crop protection. *Acta Zoologica Fennica* 173, 191-192.

Hunter, J. E. (1991). Grazing and pocket gopher abundance in a California annual grassland. *Southwestern Naturalist* 36,117-118.

Hygnstrom, S. E. & Virchow, D. R. (1994). Prairie Dogs. In S. E. Hygnstrom, R. M. Timm, & G. E. Larson (Eds.), *Prevention and Control of Wildlife Damage.* (Pp. B-85 – B-96), Lincoln, Nebraska: Cooperative Extension Division, University of Nebraska.

Hygnstrom, S. E., Timm, R. M., & Larson, G. E. (1994). *Prevention and Control of Wildlife Damage.* Lincoln, Nebraska: University of Nebraska Cooperative Extension.

Hygnstrom, S. E., VerCauteren, K. C., & Ekstein, J. D. (1996). Impacts of field-dwelling rodents on emerging field corn. In R. M. Timm & A. C. Crabb (Eds.), *Proceedings of the 17th Vertebrate Pest Conference* (Pp. 148-150), Rohnert Park, California: 17th Vertebrate Pest Conference (Mar 5-7, 1996).

Jackson, J. E. (1988). Terrestrial mammalian pests in Argentina – An overview. In A. C. Crabb & R. E. Marsh (Eds.), *Proceedings of the 13th Vertebrate Pest Conference* (Pp. 196-198), Monterey, California: 13th Vertebrate Pest Conference (Mar 1-3, 1988).

Jackson, J. J. (1994). Tree Squirrels. In S. E. Hygnstrom, R. M. Timm, & G. E. Larson (Eds.), *Prevention and Control of Wildlife Damage.* (Pp. B-171 – B-175), Lincoln, Nebraska: Cooperative Extension Division, University of Nebraska.

Jackson, W. B. (2001). Current rodenticide strategies. *International Biodeterioration and Biodegradation* 48, 127-136.

Jacob, J. (2008). Response of small rodents to manipulations of vegetation height in agro-ecosystems. *Integrative Zoology* 3, 3-10

Jacob, J., Brown, P. R., Aplin, K. P., & Singleton, G. R. (2002). Ecologically-based management of pest rodents in rice-based agro-ecosystems in Southeast Asia. In R. M.

Timm & R. H. Schmidt (Eds.), *Proceedings of the 20th Vertebrate Pest Conference* (Pp. 67-74), Reno, Nevada: 20th Vertebrate Pest Conference (Mar 4-7, 2002).

Jacob, J., Rahmini, & Sudarmaji. (2006). The impact of imposed female sterility on field populations of ricefield rats (*Rattus argentiventer*). *Agriculture, Ecosystems & Environment* 115, 281-284.

Jahn, G. C., Solieng, M., Cox, P. G., & Nel, C. (1999). Farmer participatory research on rat management in Cambodia. In G. R. Singleton, L. A. Hinds, H. Leirs, and Z. Zhang (Eds.), *Ecologically-based management of rodent pests.* (Pp. 358-371), Canberra, Australia: Australian Centre for International Agricultural Research.

Jaksic, F. M. & Lima, M. (2003). Myths and facts on ratadas: Bamboo blooms, rainfall peaks and rodent outbreaks in South America. *Austral Ecology* 28, 237–251.

Joshi, R. C., Matchoc, O. R. O, Bahatan, R. G. & Dela penä, F. A. (2000). Farmers' knowledge, attitudes and practices of rice crop and pest management at Ifugao Rice Terraces, Philippines. *International Journal of Pest Management* 46, 43-48.

Kawashima, H., Bazin, M. J., & Lynch, J. M. (1997). A modelling study of world protein supply and nitrogen fertilizer demand in the 21st century. *Environmental Conservation* 24, 50-57.

King, C. M. (Ed.) (2005). *The Handbook of New Zealand Mammals Second edition.* Aukland, New Zealand: Oxford University Press. 600 pp.

Khamphoukeo, K., Douangboupha, B., Aplin, K. A., & Singleton, G.R. (2003). Pest and non pest rodents in the upland agricultural landscape of Laos: a progress report. In G. R. Singleton, L. A. Hinds, C. J. Krebs, & D. M. Spratt (eds.), *Rats, Mice and People: Rodent Biology and Management.* (Pp. 284-289), Canberra, Australia: ACIAR.

Khoo, K. C., Lam, Y. M., Teoh, C. H., Lim, W. H., & Mohamad, B. M. (1982). *Rodent Pests of Agricultural Crops in Malaysia.* Kuala Lumpur, Malaysia: Malaysian Plant Protection Society. 69 pp.

Khoprasert, Y., Promkerd, P., Ginting, S., & Jakel, T. (2008). Combination of biological rodent control using *Sarcocystis singaporensis* with anticoagulant rodenticides. In E. N. Burton & P. V. Williams (Eds.), *Crop protection research advances.* (Pp. 145-161), New York, New York: Nova Science Publishers, Inc.

Kogan, M. & Lattin, J. D. (1999). Agricultural systems as ecosystems. In J. R. Ruberson (Ed.), *Handbook of Pest Management.* (Pp. 1-33), New York: Marcel Dekker, Inc.

Korpimäki, E., Brown, P. R., Jacob, J., & Pech, R. P. (2004). The puzzles of population cycles and outbreaks of small mammals solved? *BioScience* 54, 1071-1079.

Krebs, C. J. (1999). Current paradigms of rodent population dynamics – What are we missing? In G. R. Singleton, L. A. Hinds, H. Leirs, & Z. Zhang (Eds.), *Ecologically-based management of rodent pests.* (Pp. 33-48), Canberra, Australia: ACIAR Monograph No. 59.

Krebs, C. J. (1996). Population cycles revisited. *Journal of Mammalogy* 77, 8-24.

Krebs, C., Kenney, A., Singleton, G., Mutze, G., Pech, R., Brown, P., & Davis, S. (2004). Can outbreaks of house mice in south-eastern Australia be predicted by weather models? *Wildlife Research* 31, 465-474.

Lan, L. P., Aplin, K. P., Hung, M. N., Quoc, N. V., Chien, H. V., Sang, N. D., & Singleton, G. R. (2003). Rodent communities and historical trends in rodent damage in the Mekong Delta of Vietnam: Establishing an ecological basis for effective pest management. In G. R. Singleton, L. A. Hinds, C. J. Krebs, & D. M. Spratt (Eds.), *Rats, Mice and People:*

Rodent Biology and Management. (Pp. 290-296), Canberra, Australia: ACIAR Monograph No. 96.

Leirs, H. (2003). Management of rodents in crops: the pied piper and his orchestra. In G. R. Singleton, L. A. Hinds, C. J. Krebs, and D. M. Spratt (Eds.), *Rats, mice and people: Rodent biology and management.* (Pp. 183-190), Canberra, Australia: Australian Centre for International Agricultural Research.

Leirs, H. (1999). Populations of African rodents: models and the real world. In G. R. Singleton, L. A. Hinds, H. Leirs, & Z. Zhang (Eds.), *Ecologically-based Management of Rodent Pests.* (Pp. 388-408), Canberra, Australia: Australian Centre for International Agricultural Research.

Leirs, H. & Schockaert, E. (1997). *Rodent Biology and Integrated Pest Management in Africa: Belgian Journal of Zoology.* From the proceedings of the International Workshop held in Morogoro (Vol. 127 supp. 1, 1-180), Tanzania, Africa: the International Workshop held in Morogoro (October 21-25, 1996).

Leirs, H., & Singleton, G. (2006). Parasites and pest population management. In S. Morand, B. Krasnov, R. Poulin (eds.), *Mammals and macroparasites: from evolutionary ecology to management.* (Pp. 565-591), Tokyo, Japan: Springer-Verlag.

Leirs, H., Singleton, G. R., & Hinds, L. A. (1999). Ecologically-based rodent management in developing countries: Where to now? In G. R. Singleton, L. A. Hinds, H. Leirs, and Z. Zhang (Eds.), *Ecologically-based management of rodent pests.* (Pp. 477-484), Canberra, Australia: Australian Centre for International Agricultural Research.

Leirs, H., Verhagen, R., Verheyen, W., Mwanjabe, P., & Mbise, T. (1996). Forecasting rodent outbreaks in Africa: An ecological basis for *Mastomys* control in Tanzania. *Journal of Applied Ecology* 33, 937-943.

Leung, L. K. P., Singleton, G. R., Sudarmaji & Rahmini. (1999). Ecologically-based population management of the rice-field rat in Indonesia. In G. R. Singleton, L. A. Hinds, H. Leirs, and Z. Zhang (Eds.), *Ecologically-based management of rodent pests.* (Pp. 305-319), Canberra, Australia: Australian Centre for International Agricultural Research.

Lund, M. (1988). Rodent problems in Europe. In I. Prakash (Ed.), *Rodent pest management.* (Pp. 29-33), Boca Raton, Florida: CRC Press, Inc.

Lund, M. (1984). Vertebrate pest problems in West Europe, assessment of damage and methods of control. In A. C. Dubock (Ed.), *The Organization and Practice of Vertebrate Pest Control.* (Pp. 1-5), Surrey, U. K.: ICI.

Macdonald, D. W., Mathews, F., & Berdoy, M. (1999). The behaviour and ecology of *Rattus norvegicus*: From opportunism to kamikaze tendencies. In G. R. Singleton, L. A. Hinds, H. Leirs, and Z. Zhang (Eds.), *Ecologically-based management of rodent pests.* (Pp. 49-80), Canberra, Australia: Australian Centre for International Agricultural Research.

Makundi, R. H., Oguge, N. O., & Mwanjabe, P. S. (1999). Rodent pest management in east Africa – an ecological approach. In G. R. Singleton, L. A. Hinds, H. Leirs, and Z. Zhang (Eds.), *Ecologically-based management of rodent pests.* (Pp. 460-476), Canberra, Australia: Australian Centre for International Agricultural Research.

Marsh, R. E. (1994). Current (1994) ground squirrel control practices in California. In W. S. Halverson & A. C. Crabb (Eds.), *Proceedings of the 16th Vertebrate Pest Conference* (Pp. 61-65), Santa Clara, California: 16th Vertebrate Pest Conference (Mar 1-3, 1994).

Marsh, R. E. (1984). On overview of rodent and rabbit pests of agricultural crops in the United States. In A. C. Dubock (Ed.), *The Organization and Practice of Vertebrate Pest Control*. (Pp. 57-60), Surrey, U. K.: ICI.

Marsh, R. E. (1997). *Pocket Gopher Traps: A Collectors Manual*. Davis, California: Department of Wildlife, Fish, and Conservation Biology, University of California. 305 pp.

Marsh, R. E. (1992). Reflections on current (1992) pocket gopher control in California. In J. E. Borrecco & R. E. Marsh (Eds.), *Proceedings of the 15th Vertebrate Pest Conference* (Pp. 289-295), Newport Beach, California: 15th Vertebrate Pest Conference (Mar 3-5, 1992).

Marsh, R. E. (1988). Rodent problems on the North American continent. In I. Prakash (Ed.), *Rodent pest management*. (Pp. 1-11), Boca Raton, Florida: CRC Press, Inc.

Marsh, R. E., Koehler, A. E., & Salmon, T. P. (1990). Exclusionary methods and materials to protect plants from pest mammals – A review. In L. R. Davis & R. E. Marsh (Eds.), *Proceedings of the 14th Vertebrate Pest Conference* (Pp. 174-180), Sacramento, California: 14th Vertebrate Pest Conference (Mar 6-8, 1990).

Mason, J. R. (1998). Mammal repellents: options and considerations for development. In R. O. Baker & A. C. Crabb (Eds.), *Proceedings of the 18th Vertebrate Pest Conference* (Pp. 325-329), Costa Mesa, California: 18th Vertebrate Pest Conference (Mar 2-5, 1998).

Massawe, A. W., Leirs, H., Rwamugira, W. P., & Makundi, R. H. (2003). Effect of land preparation methods on spatial distribution of rodents in crop fields. In G. R. Singleton, L. A. Hinds, C. J. Krebs, and D. M. Spratt (Eds.), *Rats, mice and people: Rodent biology and management.* (Pp. 229-232), Canberra, Australia: Australian Centre for International Agricultural Research.

Massawe, A. W., Rwanmugira, W., Leirs, H., Makundi, R. H., Mulungu, L., Ngowo, V., & Machang'u, R. (2008). Soil type limits population abundance of rodents in crop fields: Case study of the multimammate rat [*Mastomys natalensis (Smith)*], 1834 in Tanzania. *Integrative Zoology* 3, 27-30.

Mathur, R. P. (1997). Effectiveness of various rodent control measures in cereal crops and plantations in India. In H. Leirs & E. Schockaert (Eds.), *Rodent biology and integrated pest management in Africa: Belgian journal of zoology*. From the Proceedings of the International Workshop held in Morogoro (Vol. 127. supp. 1, 137-144), Tanzania, Africa: the International Workshop held in Morogoro (October 21-25, 1996).

Matthews, G. A. (1996). The importance of scouting in cotton IPM. *Crop Protection* 15, 369-374.

McComb, W. C. (1992). Modeling systems for prescription development. In H. C. Black (Ed.), *Sivercultural approaches to animal damage management in Pacific Northwest Forests*. (Pp. 377-393), Portland, Oregon: Department of Agriculture, Forest Service, Pacific Northwest Research Station.

McGlinchy, A. (1999). *PestCalc: The Pest Control Calculator Users Guide*. Lincoln, New Zealand: Manaaki Whenua Lancare Research. 15 pp.

McLeod, S. R., Saunders, G., Twigg, L. E., Arthur, A. D., Ramsey, D., & Hinds, L. A. (2007). Prospects for the future: Is there a role for virally vectored immunocontraception in vertebrate pest management? *Wildlife Research* 34, 555-566.

Meehan, A. P. (1984). *Rats and Mice Their Biology and Control*. W. Sussex, Felcourt, East Grinstead: Rentokil Limited. 383 pp.

Meerburg, B. G., Singleton, G. R., & Kijlstra, A. (2009a). Rodent-borne diseases and their risks for public health. *Critical Reviews in Microbiology* 35, 221-270.

Meerburg, B. G., Singleton, G. R., & Leirs, H. (2009b). The Year of the Rat ends – time to fight hunger. *Pest Management Science* 65, 351-352.

Miller, L., Johns, B., & Elias, D. (1998). Immunocontraception as a wildlife management tool: some perspectives. *Wildlife Society Bulletin* 26, 237-243.

Miller, R.W., Stuart, A.M., Joshi, R.C., Banks, P.B., & Singleton, G. R. (2008). Biology and management of rodent communities in complex agroecosystems – rice terraces. In G. R. Singleton, R. C. Joshi & L.S. Sebastian (Eds.), *Philippine Rats: Ecology and Management.* (Pp. 25-36), Science City of Muñoz, Neuva Ecija: Philippine Rice Research Institute, Los Banos.

Mills, J. N., Ellis, B. A., McKee, K. T., Maiztegui, J. I., & Childs, J. E. (1991). Habitat associations and relative densities of rodent populations in cultivated areas of Central Argentina. *Journal of Mammalogy* 72, 470-479.

Miño, M. H., Cavia, R., Villafañe, I. E. G., Bilenca, D. N., & Busch, M. (2007). Seasonal abundant and distribution among habitats of small rodents on poultry farms. A contribution for their control. *International Journal of Pest Management* 53(4), 311-316.

Mohammad, A. H. H., Zaghloul, T. M., & Zakaria, M (Eds.). (1983). *Proceedings of the First Symposium on Recent Advances in Rodent Control* (Pp. 1-225), Kuwait: First Symposium on Recent Advances in Rodent Control, Ministry of Public Health (Feb 8-11, 1982).

Moran, S. (2008). Rodent management in animal farms by anticoagulant rodenticides. In E. N. Burton & P. V. Williams (Eds.), *Crop protection research advances.* (Pp. 95-117), New York, New York: Nova Science Publishers, Inc.

Moran, S. & Keidar, H. (1993). Checklist of vertebrate damage to agriculture in Israel. *Crop Protection* 12, 173-182

Moser, B. W. & Witmer, G. W. (2000). The effects of elk and cattle foraging on the vegetation, birds, and small mammals of the Bridge Creek Wildlife Area, Oregon. *International Biodeterioration and Biodegradation* 45, 151-157.

Murphy, E. C., Clapperton, B.K., Bradfield, P. M. F., & Speed, H. J. (1998). Effects of rat-poisoning operations on abundance and diet of mustelids in New Zealand podocarp forests. *New Zealand Journal of Zoology* 25, 315-328.

Mwanjabe, P. S. & Leirs, H. (1997). An early warning system for IPM-based rodent control in smallholder farming systems in Tanzania. In H. Leirs & E. Schockaert (Eds.), *Rodent biology and integrated pest management in Africa: Belgian journal of zoology.* From the Proceedings of the International Workshop held in Morogoro (Vol. 127. supp. 1, 49-58), Tanzania, Africa: the International Workshop held in Morogoro (October 21-25, 1996).

Mwanjabe, P. S., Sirima, F. B., & Lusingu, J. (2002). Crop losses due to outbreaks of [*Mastomys natalensis* (Smith, 1834)] Muridae, Rodentia, in the Lindi Region of Tanzania. *International Biodeterioration & Biodegradation* 49, 133-137.

Naiman, R. J., Johnston, C. A., & Kelley, J. C. (1988). Alteration of North American Streams by Beaver. *BioScience* 38, 753-762.

Ngowo, V., Lodal, J., Mulungu, L. S., Makundi, R. H., Massawe, A. W., & Leirs, H. (2003). Evaluation of thiram and cinnamamide as potential repellents against maize-seed depredation by the Multimammate Rat, (*Mastomys natalensis*), in Tanzania. In G. R. Singleton, L. A. Hinds, C. J. Krebs, and D. M. Spratt (Eds.), *Rats, mice and people:*

Rodent biology and management. (Pp. 260-261), Canberra, Australia: Australian Centre for International Agricultural Research.

Nolte, D. L. (1996). Mongolian rangelands: Rodent problems and approaches to alleviate damage. In R. M. Timm & A. C. Crabb (Eds.), *Proceedings of the 17th Vertebrate Pest Conference* (Pp. 143-147), Rohnert Park, California: 17th Vertebrate Pest Conference (Mar 5-7, 1996).

O'Brien, J. M. (1994). Voles. In S. E. Hygnstrom, R. M. Timm, & G. E. Larson (Eds.), *Prevention and Control of Wildlife Damage.* (Pp. B-177 – B-182), Lincoln, Nebraska: Cooperative Extension Division, University of Nebraska.

Oguge, N., Ndung'u, D, & Okemo, P. (1997). Effects of Neem Plant (Azadirachta Indica Juss, meliaceae) products on maize grain consumption by three common rodent pests in Kenya. In H. Leirs & E. Schockaert (Eds.), *Rodent biology and integrated pest management in Africa: Belgian journal of zoology.* From the Proceedings of the International Workshop held in Morogoro (Vol. 127. supp. 1, 129-135), Tanzania, Africa: the International Workshop held in Morogoro (October 21-25, 1996).

Ojwang, D. O. & Oguge, N. O. (2003). Testing a biological control program for rodent management in a maize cropping system in Kenya. In G. R. Singleton, L. A. Hinds, C. J. Krebs, and D. M. Spratt (Eds.), *Rats, mice and people: Rodent biology and management.* (Pp. 251-253), Canberra, Australia: Australian Centre for International Agricultural Research.

Olsen, P. (1998). *Australia's Pest Animals.* Canberra, Australia: Bureau of Resource Sciences and Kangaroo Press. 160 pp.

Painter, J. A., Mølbak, K., Sonne-Hansen, J., Barrett, T., Wells, J. G., & Tauxe, R. V. (2004). Salmonella-based rodenticides and public health. *Emerging Infectious Diseases* 10(6), 985-987.

Pearson, O. P. (1963). History of two local outbreaks of feral house mice. *Ecology* 44(3), 540-549.

Pech, R. P. (2000). Biological control of vertebrate pests. In T. P. Salmon & A. C. Crabb (Eds.), *Proceedings of the 19th Vertebrate Pest Conference* (Pp. 206-211), San Diego, California: 19th Vertebrate Pest Conference (Mar 6-9, 2000).

Pech, R. P., Hood, G. M., Singleton, G. R., Salmon, E., Forrester, R. I., & Brown, P. R. (1999). Models for predicting plagues of house mice (*Mus domesticus*) in Australia. In G. R. Singleton, L. A. Hinds, H. Leirs, & Z. Zhang (Eds.), *Ecologically-based management of rodent pests.* (Pp. 81-112), Canberra, Australia: ACIAR Monograph No. 59.

Pelton, M. R., Fribourg, H. A., Laundre, J. W. & Reynolds, T. W. (1991). Preliminary assessment of small wild mammal populations in tall fescue habitats. *Tennessee Farm and Home Science* 160, 68-71.

Pelz, H. (2003). Current approaches towards environmentally benign prevention of vole damage in Europe. In G. R. Singleton, L. A. Hinds, C. J. Krebs, and D. M. Spratt (Eds.), *Rats, mice and people: Rodent biology and management.* (Pp. 233-237), Canberra, Australia: Australian Centre for International Agricultural Research.

Pelz, H. J. (1987). Damage by wood mice (*Apodemus sylvaticus*) to sugarbeet seeds in relation to food availability. In C. G. J. Richards & T. Y. Ku (Eds.), *Control of Mammal Pests.* (Pp. 5-14), London, New York and Philadelphia: Taylor & Francis.

Pelz, H. (2007). Spread of resistance to anticoagulant rodenticides in Germany. *International Journal of Pest Management* 53(4), 299-302.

Posamentier, H. (1997). Communication in national rodent management programmes. In H. Leirs & E. Schockaert (Eds.), *Rodent biology and integrated pest management in Africa: Belgian journal of zoology*. From the Proceedings of the International Workshop held in Morogoro (Vol. 127. supp. 1, 129-135), Tanzania, Africa: the International Workshop held in Morogoro (October 21-25, 1996).

Posamentier, H. & Elsen, A. V. (1984). *Rodent Pests their Biology and Control in Bangladesh*. Mymensingh, Bangladesh: Bangladesh Agricultural University. 110 pp.

Prakash, I. (1988). *Rodent Pest Management*. Boca Raton, Florida: CRC Press, Inc. 480 pp.

Prakash, I. (1984). Vertebrate pest problems in India. In A. C. Dubock (Ed.), *The Organization and Practice of Vertebrate Pest Control*. (Pp. 29-35), Surrey, U. K.: ICI.

Prakash, I., & Ghosh, P. K. (1992). *Rodents in Indian Agriculture* (Volume 1). Jodhpur, India: Pawan Kumar Scientific Publishers. 707 pp.

Prakash, I. & Mathur, R. P. (1988). Rodent problems in Asia. In I. Prakash (Ed.), *Rodent pest management*. (Pp. 67-84), Boca Raton, Florida: CRC Press, Inc.

Proctor, D. L. (1994). *Grain Storage Techniques*. Rome, Italy: Food and Agriculture Organization of the United Nations Bulletin 109. 277 pp.

Putman, R. J. (1989). *Mammals as Pests*. London & New York: Chapman and Hall. 271 pp.

Quick, G. R. (Ed.). (1991). *Rodents and rice: Report and Proceedings of an Expert Panel Meeting on Rice Rodent Control*. Los Baños, Philippines: International Rice Research Institute. 132 pp.

Quy, R. J., Watkins, R. W., Lambert, M. S, & Cowan, D. P. (2009). Validating census methods to measure changes in house mouse populations. *Pest Management Science* 65, 275–282.

Ramsay, D. S. L. & Wilson, J. C. (2000). Towards ecologically based baiting strategies for rodents in agricultural systems. *International Biodeterioration & Biodegradation* 45, 183-197.

Rao, A. M. K. M. (2003). Rodent problems in India and strategies for their management. In G. R. Singleton, L. A. Hinds, C. J. Krebs, and D. M. Spratt (Eds.), *Rats, mice and people: Rodent biology and management*. (Pp. 203-212), Canberra, Australia: Australian Centre for International Agricultural Research.

Redwood, A. J., Smith, L. M., Lloyd, M. L., Hinds, L. A., Hardy, C. M., & Shellam, G. R. (2007). Prospects for virally vectored immunocontraception in the control of wild house mice (Mus domesticus). *Wildlife Research* 34, 530-539.

Richards, C. G. J. (1989). The pest status of rodents in the United Kingdom. In R. J. Putman (Ed.), *Mammals as pests*. (Pp. 21-33), New York, New York: Chapman and Hall.

Richards, C. G. J. & Ku, T. Y. (1987). *Control of Mammal Pests*. London, New York and Philadelphia: Taylor & Francis. 406 pp.

Roberts, T. J. (1981). *Hand Book of Vertebrate Pest Control in Pakistan*. Compiled by the Vertebrate Pest Control Centre. Islamabad, Pakistan: PanGraphics Ltd. 237 pp.

Roy, S., Smith, G. C., & Russell, J. C. (2009). The eradication of invasive mammal species: Can adaptive resource management fill the gaps in our knowledge? *Human-Wildlife Conflicts* 3(1), 30-40.

Salatti, C., Woolhouse, A., & Vandenbergh, J. (1995). The use of odors to induce avoidance behavior in pine voles. In M. King (Ed.), *Proceedings of the 6th Eastern Wildlife Damage Control Conference* (Pp. 149-151), Ashville, North Carolina: the 6th Eastern Wildlife Damage Control Conference (Oct 3-6, 1993).

Salmon, T. P. & Licklitter, R. E. (1983). Comparisons between vertebrate pest control materials: essential considerations. In D. E. Kaukeinen (Ed.), *Vertebrate Pest Control and Management Materials, fourth symposium*. (Pp. 5-19), Philadelphia, PA: American Society for Testing and Materials.

Schiller, J. M., Boupha, B. D., & Bounnaphol, O. (1999). Rodents in agriculture in the Lao PDR—a problem with an unknown future. In G. R. Singleton, L. A. Hinds, H. Leirs, and Z. Zhang (Eds.), *Ecologically-based management of rodent pests.* (Pp. 372-387), Canberra, Australia: Australian Centre for International Agricultural Research.

Seeley, R. R. & Reynolds, T. D. (1989). Effect of indomethacin-treated wheat on a wild population of montane voles. *Great Basin Naturalist* 49, 556-561.

Sheffield, L. M., Crait, J. R., Edge, W. D., & Wang, G. (2001). Response of American kestrels and gray-tailed voles to vegetation height and supplemental perches. *Canadian Journal of Zoology*, 79(3), 380-385.

Shumake, S. A., Kolz, A. L., Reidinger, R. F., & Fall, M. W. (1979). Evaluation of nonlethal electrical barriers for crop protection against rodent damage. In J. R. Beck (Ed.), *Vertebrate Pest Control and Management Materials, proceedings from the second symposium sponsored by American Society for Testing and Materials.* (Pp. 29-38), Philadelphia, PA: American Society for Testing and Materials.

Singla, N., Parshad, V. R., & Singla, L. D. (2003). Potential of (*Trypanosoma evansi*) as a biocide of rodent pests. In G. R. Singleton, L. A. Hinds, C. J. Krebs, & D. M. Spratt (Eds.), *Rats, Mice and People: Rodent Biology and Management*. (Pp. 43-46), Canberra, Australia: ACIAR Monograph No. 96.

Singleton, G. R. (1997). Integrated management of rodents: A southeast Asian and Australian perspective. *Belgian Journal of Zoology* 127, 157-169.

Singleton, G. R. (1989). Population dynamics of an outbreak of house mice (*Mus domesticus*) in the mallee wheatlands of Australia - hypothesis of plague formation. *Journal of Zoology (London)* 219, 495-515.

Singleton, G. R. (2003). *Impacts of rodents on rice production in Asia.* Los Baños, Philippines: IRRI Discussion Paper Series No. 45, International Rice Research Institute. 30 pp.

Singleton, G. R., Brown, P. R., Jacob, J., Aplin, K. P., & Sdarmaji. (2007a). Unwanted and unintended effects of culling: A case for ecologically-based rodent management. *Integrative Zoology* 2, 247-259.

Singleton, G.R., Colin, R., Tann, C.R., & Krebs, C.J. (2007b). Landscape ecology of house mouse outbreaks in southeastern Australia. *Journal of Applied Ecology* 44, 644-652.

Singleton, G. R., Hinds, L. A., Krebs, C. J., & Spratt, D. M. (2003). *Rats, Mice and People: Rodent Biology and Management*. Canberra, Australia: Clarus Design. 564 pp.

Singleton G. R., Joshi, R. C. & Sebastian, L. S. (Eds.). (2008). *Philippine Rats: Ecology and Management*. Science City of Muñoz, Neuva Ecija: Philippine Rice Research Institute. 215 pp

Singleton, G. R., Leirs, H., Hinds, L. A., & Zhang, Z. (1999). Ecologically-based management of rodent pests – re-evaluating our approach to an old problem. In G. R. Singleton, L. A. Hinds, H. Leirs, and Z. Zhang (Eds.), *Ecologically-based management of rodent pests.* (Pp. 17-29), Canberra, Australia: Australian Centre for International Agricultural Research.

Singleton, G. R., Sudarmaji, Jacob, J., & Krebs, C. J. (2005). Integrated management to reduce rodent damage to lowland rice crops in Indonesia. *Agriculture, Ecosystems and Environment* 107, 75-82.

Singleton, G. R., Sudarmaji, & Suriapermana, S. (1998). An experimental field study to evaluate a trap-barrier system and fumigation for controlling the rice rat in rice crops in West Java. *Crop Protection* 17, 55-64.

Smythe, W. R. (1986). Rodent control in East Africa. In T. P. Salmon (Ed.), *Proceedings of the 12th Vertebrate Pest Conference* (Pp. 188-190), San Diego, California: 12th Vertebrate Pest Conference (Mar 4-6, 1986).

Stenseth, N. C., Leirs, H., Skonhoft, A., Davis, S. A., Pech, R. P., Andreassen, H. P., Singleton, G. R., Lima, M., Machang'u, R. S., Maundi, R. H., Zhang, Z., Brown, P. R., Shi, D., Wan, X. (2003). Mice, rats, and people: the bio-economics of agricultural rodent pests. *Front Ecology Environment* 1(7), 367-375.

Sterner, R. T. (2008). Reducing the uncertainty of IPM economics. In E. N. Burton & P. V. Williams (Eds.), *Crop protection research advances*. (Pp. 163-181), New York, New York: Nova Science Publishers, Inc.

Sterner, R. T. (2002). Spreadsheets, response surfaces, and intervention decisions in wildlife damage management. In L. Clark, J. Hone, J. A. Shivik, R. A. Watkins, K. C. VerCauteren, & J. K. Yoder (Eds.), *Human Conflicts with Wildlife: Economic Considerations*, from the proceedings from the Third NWRC Special Symposium (Pp. 42-47), Fort Collins, Colorado: Third NWRC Special Symposium (August 1-3, 2002).

Stuart, A. M., Prescott, C. V. & Singleton, G. R. (2008). Biology and management of rodent communities in complex agroecosystems – lowlands. In G. R. Singleton, R. C. Joshi & L.S. Sebastian (Eds.), *Philippine Rats: Ecology and Management.* (Pp. 37-55), Science City of Muñoz, Neuva Ecija: Philippine Rice Research Institute, Los Banos.

Stuart, A. M., Singleton, G. R., Prescott, C. V., Joshi, R. C. & Sebastian, L. S. (2007). The rodent species of the Ifugao Rice Terraces, Philippines – Target or non-target species for management? *International Journal of Pest Management* 53, 139-146.

Sullivan, T. P., Crump, D. R., & Sullivan, D. S. (1988). Use of predator odors as repellents to reduce feeding damage by herbivores IV: Northern pocket gophers. *Journal of Chemical Ecology* 14, 379-389.

Sullivan, T. P. & Klenner, W. (1993). Influences of diversionalry food on red squirrel populations and damage to crop trees in young lodgepole pine forests. *Ecological Applications* 3, 708-718.

Sullivan, T. P., Krebs, J. A., & Kluge, H. A. (1987). Survey of mammal damage to tree fruit orchards in the Okanagan Valley of British Columbia. *Northwest Science* 61, 23-31.

Sullivan, T. P., Sullivan, D. S., Hogue, E. J., Lautenschlager, R. A., & Wagner, R. G. (1998). Population dynamics of small mammals in relation to vegetation management in orchard agroecosystems: compensatory responses in abundance and biomass. *Crop Protection* 17, 1-11.

Tann, C. R., Singleton, G. R., & Coman, B. C. (1991). Diet of the house mouse (*Mus domesticus*) in the mallee wheatlands of north-western Australia. *Wildlife Research* 18, 1-12.

Taylor, K. D. (1984). Vertebrate pest problems in Africa. In A. C. Dubock (Ed.), *The organization and practice of vertebrate pest control*. (Pp. 21-28), Surrey, U. K.: ICI.

Timm, R. M. (2003). Devices for vertebrate pest control: Are they of value? In K. A. Fagerstone & G. W. Witmer (Eds.), *Proceedings of the 10th Wildlife Damage Management Conference* (Pp. 152-161), Hot Springs, Arkansas: The 10th Wildlife Damage Management Conference (Apr 6-9, 2003).

Timm, R. M. (1994a). House Mice. In S. E. Hygnstrom, R. M. Timm, & G. E. Larson (Eds.), *Prevention and Control of Wildlife Damage*. (Pp. B-31 – B-46), Lincoln, Nebraska: Cooperative Extension Division, University of Nebraska.

Timm, R. M. (1994b). Norway Rats. In S. E. Hygnstrom, R. M. Timm, & G. E. Larson (Eds.), *Prevention and Control of Wildlife Damage*. (Pp. B-105 – B-120), Lincoln, Nebraska: Cooperative Extension Division, University of Nebraska.

Tobin, M. E. (1992). Rodent damage in Hawaiian macadamia orchards. In J. E. Borrecco & R. E. Marsh (Eds.), *Proceedings of the 15th Vertebrate Pest Conference* (Pp. 272-276), Newport Beach, California: 15th Vertebrate Pest Conference (Mar 3-5, 1992).

Tomich, P. Q. (1986). *Mammals in Hawaii*. Honolulu, HI: Bishop Museum Press. 375 pp.

Truszkowski, J. (1982). The impact of the common vole on the vegetation of agroecosystems. *Acta Theriologica* 27(23), 305-345.

Tyndale-biscoe, H., & Hinds, L. A. (2007). Introduction – Virally vectored immunocontraception in Australia. *Wildlife Research* 34(7), 507-510.

United Nations Development Program (UNDP). (2009). Annual Report 2009. United Nations Development Program, Office of Communication, New York. 40 pp.

Villa-Cornejo, B. (2000). Pocket Gopher (*Orthogeomys Hispidus hispidus*) damage in sugarcane fields in the state of Veracruz, Mexico. In T. P. Salmon & A. C. Crabb (Eds.), *Proceedings of the 19th Vertebrate Pest Conference* (Pp. 358-361), San Diego, California: 19th Vertebrate Pest Conference (Mar 6-9, 2000).

Villa-Cornejo, B., Lopez-Forment, W., Cornejo, M. V., & Prescott, C. V. (1998). Not all Sigmodontine rodents in the sugarcane fields in coastal Veracruz, Mexico are pests. In R. O. Baker & A. C. Crabb (Eds.), *Proceedings of the 18th Vertebrate Pest Conference* (Pp. 236-241), Costa Mesa, California: 18th Vertebrate Pest Conference (Mar 2-5, 1998).

White, J. G., Wilson, J., & Horskins, K. (2003). Habitat manipulation and rodent damage control: Reducing rodent damage in Australian macadamia orchards. In G. R. Singleton, L. A. Hinds, C. J. Krebs, and D. M. Spratt (Eds.), *Rats, mice and people: Rodent biology and management.* (Pp. 213-216), Canberra, Australia: Australian Centre for International Agricultural Research.

Wilson, D.E. & Reeder, D.M. (Eds.). (2005). *Mammal Species of the World. A Taxonomic and Geographic Reference (3rd ed)*, Baltimore, M.D.: Johns Hopkins University Press. 2,142 pp

Witmer, G. (2007). The ecology of vertebrate pests and integrated pest management (IMP). In M. Kogan & P. Jepson (Eds.), *Perspectives in Ecological Theory and Integrated Pest Management*. (Pp. 393-410), Cambridge, UK: Cambridge University Press.

Witmer, G. W. (2005). Wildlife population monitoring: Some practical considerations. *Wildlife Research* 32, 259-263.

Witmer, G. W., Boyd, F., & Hillis-Starr, Z. (2007a). The successful eradication of introduced roof rats (*Rattus rattus*) from Buck Island using diphacinone, followed by an irruption of house mice (*Mus musculus*). *Wildlife Research* 34, 108-115.

Witmer, G., Brennan, M., Dees, D., Hoffman, B., Pusateri, F., Richardson, C., & Seery, D. (2003). Black-tailed prairie dog management in urban-suburban settings: opportunities and challenges. In the Wildlife Management Institute Publications Department (Ed.), *Transactions of the 68th North American Wildlife and Natural Resources Conference* (Pp. 209-221), Winston-Salem, North Carolina: 68th North American Wildlife and Natural Resources Conference (Mar 26-29, 2003).

Witmer, G. W. & Engeman, R. M. (2007). Subterranean rodents as pests: The case of the pocket gopher. In S. Begall, H. Burda, & C. E. Schleich (Eds.), *Subterranean Rodents: News from Underground.* (Pp. 287-299), Heidelberg, Germany: Springer-Verlag Berlin.

Witmer, G. & Eisemann, J. D. (2007). Rodenticide use in rodent management in the United States: An overview. In D. L. Nolte, W. M. Arjo, & D. Stalman (Eds.), *Proceedings of the 12th Wildlife Damage Management Conference* (Pp. 114-118), Corpus Christi, Texas: 12th Wildlife Damage Management Conference (Apr 9-12, 2007).

Witmer, G., Eisemann, J. D., & Howald, G. (2007b). The use of rodenticides for conservation efforts. In D. L. Nolte, W. M. Arjo, & D. Stalman (Eds.), *Proceedings of the 12th Wildlife Damage Management Conference* (Pp. 160-167), Corpus Christi, Texas: 12th Wildlife Damage Management Conference (Apr 9-12, 2007).

Witmer, G. W., Fall, M. W., & Fiedler, L. A. (1995). Rodent control, research, and technology transfer. In J. Bissonette & P. Krausman (Eds.), *Integrating people and wildlife for a sustainable future*, from the Proceedings of the First International Wildlife Management Congress (Pp. 693-697), Bethesda, MD: The Wildlife Society.

Witmer, G. & Fantinato, J. (2003). Management of rodents at airports. In K. A. Fagerstone & G. W. Witmer (Eds.), *Proceedings of the 10th Wildlife Damage Management Conference* (Pp. 350-358), Hot Springs, Arkansas: 10th Wildlife Damage Management Conference (Apr 6-9, 2003).

Witmer, G., Gionfriddo, J., & Pipas, M. (2008b). Evaluation of physical barriers to prevent prairie dog colony expansion. *Human-Wildlife Conflicts* 2(2), 206-211.

Witmer, G., Hakim, A., & Moser, B. (2001). Investigations of methods to reduce damage by voles. *Proceedings of the Eastern Wildlife Damage Control Conference* 9, 357-365.

Witmer, G. & Jojola, S. (2006). What's up with house mice? – A review. In R. M. Timm and J. M. O'Brien (Eds.), *Proceedings of the 22nd Vertebrate Pest Conference* (Pp. 124-130), Berkley, California: 22nd Vertebrate Pest Conference (Mar 6-9, 2006).

Witmer, G., Pipas, M., Burke, P., Rouse, D., Dees, D., & Manci, K. (2008a). Raptor use of artificial perches at natural areas, City of Fort Collins, Colorado. *The Prairie Naturalist* 40, 37-42.

Witmer, G., Sayler, R., Huggins, D., & Capelli, J. (2007c). Ecology and management of rodents in no-till agriculture in Washington, USA. *Integrative Zoology* 2, 154-164.

Witmer, G. W., Snow, N. P., Humberg, L., & Salmon, T. (2009). Vole problems, management options, and research needs in the United States. In J. Boulanger (Ed.), *Proceedings of the 13th Wildlife Damage Management Conference* (Pp. 235-249), Saratoga Springs, New York: 13th Wildlife Damage Management Conference (May 4-8, 2009).

Witmer, G. W. & VerCauteren, K. C. (2001). Understanding vole problems in direct seeding – strategies for management. In R. Veseth (Ed.), *Proceedings of the Northwest Direct Seed Cropping Systems Conference* (Pp. 104-110), Spokane, Washington: Northwest Direct Seed Conference (Jan 17-19, 2001).

Wolf, Y. (1977). The Levant vole, [*Microtus guentheri* (Danford et Alston, 1982)]. Economic Importance and Control. *EPPO Bulletin* 7, 277-281.

Wywialowski, A. P. (1998). Are wildlife-caused losses of agriculture increasing? In R. O. Baker & A. C. Crabb (Eds.), *Proceedings of the 18th Vertebrate Pest Conference* (Pp. 363-370), Costa Mesa, California: 18th Vertebrate Pest Conference (Mar 2-5, 1998).

Ylönen, H., Jacob, J., Davies, M. & Singleton, G. R. (2002). Predation risk and habitat selection of Australian house mice (*Mus domesticus*) during an incipient plague: desperate behaviour due to food depletion. *Oikos* 99, 284-289.

Zhang, M., Wang, K., Wang, Y., Guo, C., Li, B., & Hwuang, H. (2007). Recovery of a rodent community in an agro-ecosystem after flooding. *Journal of Zoology* 272, 138-147.

Zhang, Z., Chen, A., Ning, Z., & Huang, X. (1999). Rodent pest management in agricultural ecosystems in China. In G. R. Singleton, L. A. Hinds, H. Leirs, and Z. Zhang (Eds.), *Ecologically-based management of rodent pests.* (Pp. 261-284), Canberra, Australia: Australian Centre for International Agricultural Research.

Zhao, M., Liu, M., Li, D., Wan, X., Hinds, L. A., Wang, Y., & Zhang, Z. (2007). Anti-fertility effect of levonorgestrel and quinestrol in Brandt's voles (*Lasiopodomys brandtii*). *Integrative Zoology* 2, 260-268.

Zhi, D. & Cheng-Xin, W. (1984). Rodent control in China. In D. O. Clark (Ed.), *Proceedings of the 11th Vertebrate Pest Conference* (Pp. 47-53), Sacramento, California: 11th Vertebrate Pest Conference (Mar 6-8, 1984).

Zhong, W., Wang, M., & Wan, X. (1999). Ecological management of Brandt's Vole (*Microtus brandti*) in Inner Mongolia, China. In G. R. Singleton, L. A. Hinds, H. Leirs, and Z. Zhang (Eds.), *Ecologically-based management of rodent pests.* (Pp. 199-214), Canberra, Australia: Australian Centre for International Agricultural Research.

In: Agricultural Production
Editors: Felix C. Wager, pp. 39-78

ISBN: 978-1-61668-695-6

Chapter 2

SPATE IRRIGATION OF RANGELANDS: A DROUGHT MITIGATING MECHANISM

Sayyed Hamid Mesbah and Sayyed Ahang Kowsar
Fars Research Center for Agriculture and Natural Resources,
Shiraz, Islamic Republic of Iran

ABSTRACT

Transhumant pastoralism is an ecologically sound, a financially viable and a socially acceptable way of life for millions of the inhabitants of drylands. Strict observation of the carrying capacity of rangelands resulted in a sustainable management of the pastoralists' habitats for millennia. However, an unruly population growth and recurring prolonged droughts in recent years have devastated many mobile pastoralists and accelerated desertification of their domains. Provision of water, food, feed, fiber and fuelwood through spate irrigation along the pastoralists' trails may save many lives, decrease the number of environmental refugees, stem the city-ward migration, and sustain the transhumance tradition. Results of an 18-year study in a sandy desert, the previous wintering ground of nomads in southern Iran with a mean annual precipitation (MAP) of 243 mm, indicate that spate irrigation of a denuded rangeland increases its productivity both in wet (>200 mm MAP) and dry (<200 mm MAP) years. Mean usable forage yield was 490.77 and 182.53 kg ha^{-1} yr^{-1} for the spate-irrigated and control, respectively, in the wet years; this was 331.45 and 115.82 kg ha^{-1} yr^{-1} for the spate-irrigated and control, respectively, in the dry years; and 411.11(range: 136.27-721.50) and 149.18(range:78.00-350.20)kg ha^{-1} yr^{-1} for the spate-irrigated and control, respectively, during the 18-year period. Mean crown cover percentage for the mentioned treatments were 37.57 and 21.02 for the wet years; 26.97 and 17.46 for the dry years; and 32.27(range: 18.59-67.09) and 19.24 (range: 9.70-30.20) during the 18-year period. The higher yield ratio of spate-irrigated rangeland relative to control in droughts relative to wet years, 2.86 vs. 2.69, indicates the presence of a "pulse reserve" in the treated system. Moreover, contrary to the dictum of some ecologists, adapted annuals are recommended for rangeland improvement. Deposition of nutritious suspended load improved the available water holding capacity and fertility of sandy soils. Besides higher yield and better protection against wind and water erosion, the spate-irrigated land provides more nutritious feed and a longer grazing period. Furthermore, the artificial recharge of groundwater is achieved simultaneously if the spate-irrigated land is underlain with coarse-grained alluvium. Conversion of abandoned farm fields, whose productivity has been drastically impaired, to permanent grassland enhances the feed security of mobile graziers, while increases carbon sequestration, and thus mitigates the global warming.

INTRODUCTION

Mobile pastoralism, as has been demonstrated by the Iranian nomads from time immemorial, is environmentally sustainable, financially sound, and socially acceptable in desert ecosystems (Davies, 2008; Adeel et al., 2007). The nomads had learned to adapt themselves to the vagaries of climate by planning their itineraries and modifying the number of their livestock according to the forage availability. The early settlers called their new home Irana, the Land of Arians (*nomads* in Old Persian). In general, nomads are well-versed ecologists who shrewdly manage the environment; otherwise, they could not have lived that life for millennia.

"Climate, and more specifically precipitation, can be regarded as the most important single factor influencing range production" (Stoddart et al., 1975). Water shortage is the most limiting constraint in drylands. As precipitation in the nomads' habitat was inadequate for a sedentary way of life, they learned that they had to obey the dictates of nature (Tannehill, 1947). They had to make use of high mobility to adapt to the ever-changing dynamics of grasslands (Vetaas and Knudson, 2004). Moreover, their life in tents forced them to avoid exposure to extreme temperatures. Therefore, they moved between northerly summer camp sites and pastures at higher altitudes and lower, southerly sites in winter. By doing so, they utilized hundreds of thousands of square kilometers for grazing in order to maintain stocking rate numbers in equilibrium with pasture productivity. Overgrazing was considered a sacrilege (Adeel et al., 2008). Thus, it is not surprising to find out that what the Iranian nomads had learned through thousands of years of experience, only recently has been proven incontrovertible under similar environmental conditions.

It has been shown that the nomadic livestock systems are well adjusted to the ecosystems of the southern Sahel region (Breman and de Wit, 1983), and even to the Himalayan environment; continuance of transhumance within the carrying capacity of the Niti valley (Nanda Devi Biosphere Reserve buffer zone) has been advised for effective management of available resources (Nautiyal et al., 2003). Moreover, Breman and de Wit (1983) have demonstrated that the undeveloped, animal-based communities such as the southern Sahelian nomads are more energy efficient than the modern, fuel-based societies.

Nomads spread their herds evenly across the landscape, thus causing less damage to the soil (Coughenour et al., 1985). “Mobility is an ecological necessity, and the mobile pastoralism is often the best way to manage dry environments sustainably” (UNDP, 2003). On the contrary, concentration of humans and livestock into small areas degrades the soil, decreases the productive capacity of the land, and causes a decline in water resources and their quality. Sedentarization of nomad pastoralists has been an exercise in futility, and this has been concisely spelled out in a recent document by the United Nations (Adeel et al., 2009). Unfortunately, the decision-makers in drylands are neither ecologists nor heed their consultation!

Expediency lead the Government of Iran in the 1930s to decree that nomadism was an anachronism in the 20th century; those who practiced it must settle down, cultivate crops and raise domesticated animals. It was hard to rule the trekking people, scattered in far-away mountains and plains. That was how the Gareh Bygone Plain (GBP), an intermountain expanse in the Southern Zagros, a beautiful shrubland teeming with gazelles, houbara bustards, rock and turtle doves, see see partridges and many other game birds became a permanent settlement for a few clans of the Khamseh Tribe. This ecologically unsound policy initiated an accelerated process of human-induced desertification. As prime arable lands had been broken generations ago, they had to settle for what was available. Since most of these regions were not suitable for dryland farming they had to resort to irrigation. As groundwater was the only available resource, they over-exploited it.

The ill-conceived land reform in the 1960s further exacerbated the situation for the Iranian rangelands. The newly landed farmers broke the mostly sloping land not suitable for agriculture. Each farming community was allocated twice the area of its farm fields for grazing the livestock. However, in most instances the areas designed for grazing were ploughed with heavy machinery and used for cultivating crops. The national policies in the late 1970s starting to foster self-sufficiency in wheat and red meat further encouraged illegal expropriators to continue to overexploit the remaining land (Adeel et al, 2008). Thus, the *coup de grâce* was dealt to our rangelands.

Hunger, ignorance and short-sightedness, in collaboration with inappropriate technologies (ploughs and pumps), wreaked havoc on the delicate environment of the Gareh Bygone Plain. That is how this desolate, wind-swept desert, unfit even for the hardiest of nomads, was created.

Land degradation, loss of biodiversity, and above all, water shortage lead to abject poverty. En masse, cityward migration was an inevitable consequence of the three above-mentioned ill-fated policies. The hapless nomads flocked to the city for the opportunities that did not exist at home. As the environmental refugees had overstayed their welcome, the Government decided to entice them to return to their abode by providing domestic and irrigation water for them and forage for their livestock.

It should be realized that overgrazing for millennia, particularly in the 20th century in our mountainous rangelandshas, has ruined our head watersheds.Therefore, forage production in the plains, and excluding the graziers from basins that are our water towers, helps ameliorate these drastically disturbed lands; thus, it must be accepted as a watershed management priority.

Damming flush floods is usually the first solution that crosses the mind of ecologically illiterate water-resources engineers who work in mountainous regions in the dry zones. High evaporation and sedimentation rates, exorbitant costs and a long duration needed for study and construction of a dam, and above all, the subsequent environmental hazards dictated selection of another alternative, a synergistic approach for the GBP. The serendipitous discovery of a coarse alluvial fan with an estimated capacity of 100-150 million m^3 (mm^3) for water storage lead us to the planning and implementation of a spate irrigation project for rangeland improvement and the artificial recharge of groundwater (ARG) and its optimization, aquifer management (*Abkhandari* in Farsi, Kowsar, 2005). It is imperative to realize that aquifer management, a paradigm-based technology, is more than just raising the watertable. It is a budding philosophy that may change the outlook of deserts as potential living spaces. Along with alleviating desertification and its aftermath, carbon sequestration, a global concern, should also be addressed (Kowsar, 2008c).

Physical features, particularly climate, soil, and topography determine the nature of vegetation (Stoddart et al., 1975).The basins and ponds of the ARG systems function as woodlots, pasture and farm fields, too; therefore, the residual water following flood termination is consumed in evapotranspiration of indigenous or cropped plants, not explicitly in evaporation. Sedimentation, which is the nemesis of hydraulic engineers, is a great advantage in the ARG activities, because it builds a layer of nutritious soil over the system. As land degradation leads to desertification, and soil erosion is characterized as "pseudo-drought", a consequence of the loss of moisture–holding capacity (National Oceanic and Atmospheric Administration and University of Missouri-Columbia, 1979, as quoted by Wolman, 1985), soil building may be considered a desertification control mechanism and a drought avoiding activity.

The world is facing an ever-increasing demand for soil resources as the ever-growing human population needs arable land to produce food, fiber, forage and fuelwood (4Fs). According to Adeel et al. (2009) "A paradigm shift from *exploiting the land to renewing the land* is essential". Our very survival is threatened by the annual loss of more than 13 million hectares (mha) of forests (FAO, 2005)[1], and 5-6 mha of cropland (WRI, 1998). Moreover, every year, 20 mha of farmland becomes unfit for crops or buried by urban and infrastructure development (UNEP, 2007). Therefore, spate irrigation of mostly degraded rangelands and abandoned farm fields enhances the soil quality while expands the arable land.

The intrinsic very low water holding capacity and infertility renders sand and gravel, on which the ARG systems are constructed, unsuitable for cropping. Spate irrigation of shelterbelts and windbreaks erected around farm fields and pastures established on such terrains provides shade for the ruminating livestock, supplies fuel wood, builds soil, mitigates dust storms, sequesters carbon and ameliorates the weather; this also reduces the nonrenewable energy spent in pumping water, and in heating and cooling the living spaces (Kowsar, 2008b). Last but by no means least, flood damage mitigation is the most obvious outcome of the ARG activities. In this paper we discuss only those items that are a matter of concern to the sedentarized nomads. Carbon sequestration, which is a worldwide concern, will be reported separately.

FLOODWATER SPREADING CONCEPTS

1. While the mean annual rainfall in drylands is very low, its variability is extremely high (erratic in distribution and frequency). The chance of receiving the desired rainfall at the expected time is meager, indeed. Therefore, water, the most limiting constraint in arid zones, is also the most precious commodity, and flood is its largest supplier in drylands. Although floodwater is a renewable capital, its use must be optimized.

[1] . FAO, 2005; WRI, 1998; and UNEP, 2007 have been quoted from Adeel et al., 2009.

2. Available water capacity (AWC) is the most important direct driver of ecological sustainability in drylands. Soil *texture* and *depth* are the two major determinants of AWC.
3. Water, an erosion agent, may become a land renovator and soil builder by transporting to and depositing nutritious sediment on the slightly sloping, drastically disturbed lands, thus improving their texture and increasing their depth and enhancing their AWC. This process also modifies the topography, stabilizes the moving sands and deprives the wind of the erodible material at its source.
4. High evaporation rate from surface waters, rapid siltation, inundation of productive lands, dwellings and archeological sites, forced migration of the inhabitants of the inundated area and construction sites, threat to biodiversity, reservoir leakage, earthquakes and other related environmental hazards, infringement of the riparian rights, decrease in the natural recharge of the downstream aquifers, and the very high costs and relatively long time needed for their construction, make large dams the most inappropriate technology in drylands. Furthermore, such schemes do not benefit inhabitants of the runoff-producing headwater catchments. They not only lose most of the water that Nature bestows upon them, but also the surface soil, the life-giving substance that nothing can replace.
5. Turbid floodwater should be harnessed to build soil and produce virtual water through spate irrigation, and/or it should be stored in aquifers by employing the artificial recharge of groundwater methods and used commensurate with needs. Doing so would help to mitigate flooding hazards, too (Kowsar, 2008a).

MATERIALS AND METHODS

Geological Setting

Subduction of the Afro-Arabian Plate under the Iranian Plate during the Upper Cretaceous, which continues to this day, has resulted in the formation of the Zagros Mountains and the rising of the Iranian Plateau. This rise that is about 5 mm per year is mainly due to the opening of the Red Sea, and the 10-20 mm annual migration of the Arabian Homocline (Farhoudi, 1987). The tectonic movements of the Zagros Mountains during the Mio- Pliocene time in the Agha Jari Formation (AJF), has made the 192 km^2 Bisheh Zard Basin, east of the GBP, an ideal runoff-producing watershed, and the northwest-southeast trending valley to its west, a large depression in which a vast amount of coarse-grained alluvium could be stored. The tremendous lateral pressure of the Afro-Arabian Plate to the several thousand meter thick carbonate deposits of the Zagros Mountains caused breakage of rocks and formation of talus and colluvium to be transported by the flowing water and form the alluvial aquifers during the Quaternary period.

The AJF consists of calcareous sandstones, siltstones, and red marls; therefore, it has formed a rather impervious bedrock. Furthermore, the Pleistocene Bakhtyari Formation, which consists of a conglomerate with calcareous and cherty cobbles and pebbles, covers the AJF unconformably (James and Wynd, 1965). The westward flowing Bisheh Zard Ephemeral River, which drains the Bisheh Zard Basin, has deposited the alluvium derived from these 2 formations; it contains 50-70% cobbles and pebbles with the specific yield of about 10%. The aquifer formed in this deposit may hold 100-150 mm^3 of water. The area over this aquifer is called the Gareh Bygone Plain.

The coordinates of the center of this plain are: 28° 35′ N, 53° 53′ E; 1140 m above mean sea level (Fig.1). The Tchah Qootch Ephemeral River that drains a 178 km^2 watershed adjacent to the Bisheh Zard Basin also supplies the ARG systems east of the Bisheh Zard Ephemeral River. The Gehrab Ephemeral River enters the GBP through its southern corner. The estimated capacity of the aquifer recharged by these 2 streams is 100 mm^3. The specific yield of this aquifer is about 8%. The gauging station of the Bisheh Zard Ephemeral River has been operative since December 2002; the other 2 streams are not gauged.

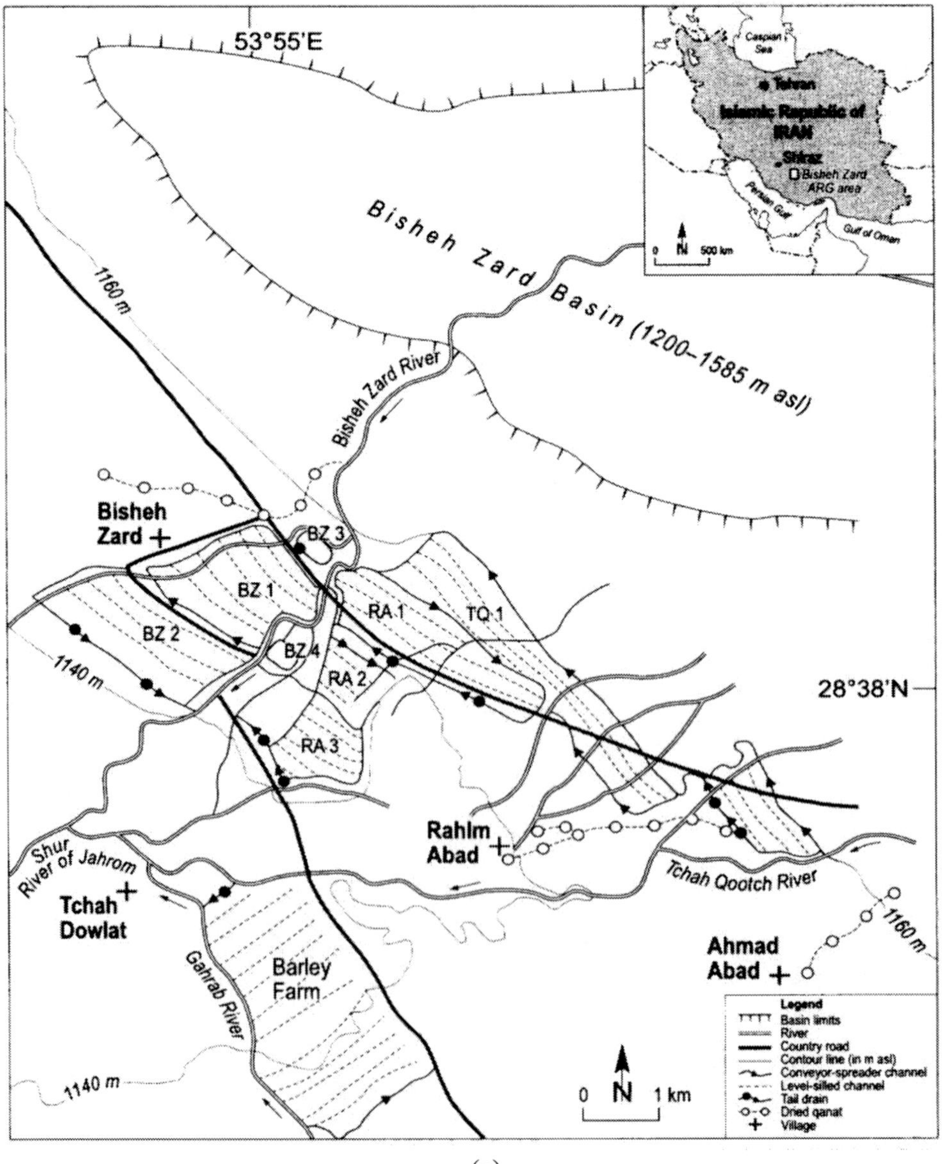

(a)

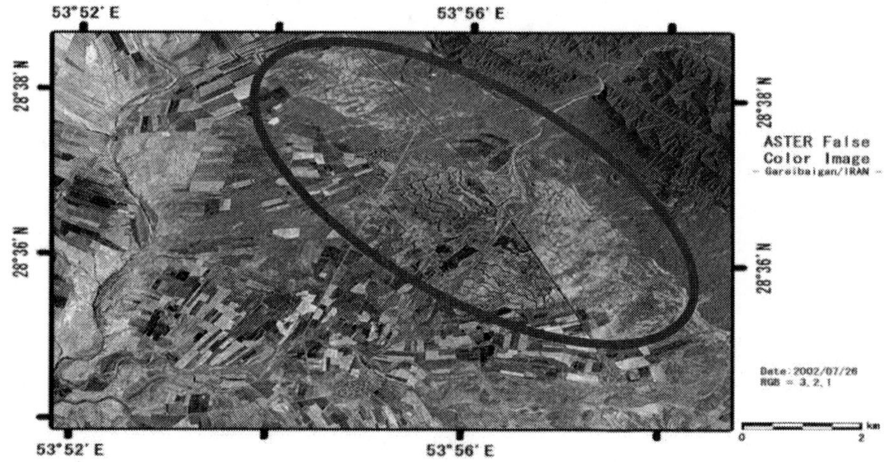

(b)

Figure 1(a). Map illustrating the study area of the Gareh Bygone Plain, Islamic Republic of Iran. (b) Satellite imagery of the Gareh Bygone Plain. The artificial recharge of groundwater systems are delineated by the red ellipse.

Climate

The Gareh Bygone Plain climate is characterized by a very low and highly variable winter rainfall, which is mostly brought in by the clouds originating over the Mediterranean Sea and summer dryness. On some rare events we welcome the clouds formed over the Sudan and even Central Asia. Convective downpours in the summer are most welcome as they save the livestock long distance walks to water holes. The floodwater that remains in the ARG systems channels serves both animals and men, as it is of a better quality than the groundwater. The mean annual rainfall and the Class A pan evaporation are 243 and 3200 mm, respectively (Pooladian and Kowsar, 2000). The mean annual temperature is 19°C, and the recorded absolute maximum and minimum temperatures are 43°C (July) and -7°C (January).

Hot, dry winds, which usually blow from the southwest during late spring and summer, raise the temperature to around 50°C in the shade. Moist winds sometimes blow from the south in the fall and winter. These winds precede heavy rains in cloudy days (Kowsar, 1991).

The time of flood occurrence and the harvested volume of floodwater are the most important factors in the productivity of spate-irrigated rangelands. Therefore, relying exclusively on the amount of precipitation in designing the ARG systems is hazardous, as only heavy showers or prolonged rainfalls produce runoff. However, relatively impermeable catchments may produce runoff with rains of low intensity; a 5 mm rainfall on the rocky Bisheh Zard Basin in the Gareh Bygone Plain leads to run-off if it occurs in less than one hour.

As both recurrent and prolonged droughts are common in drylands, differentiating between "dry" and "wet" years is therefore subjective. Since the rain is the only water that the control plots receive, we had no choice but to differentiate between the "dry" and "wet" years by arbitrarily assigning the 200 mm annual rainfall as the borderline separating the 2 conditions (Table 1). The daily rainfall for the period 24 November 1991 to 13 April 2009 is presented in Table 2. Table 3 depicts the flooding events during the study period. Please note that the hydrometry station on the Bisheh Zard Ephemeral River became operative in December 2002.

Table 1. Annual Precipitation (23 September-22 September) of 1991 to 2009 for the Gareh Bygone Plain.

No	Year	Precipitation, mm	Condition*
1	1991-92	273.5	Wet
2	1992-93	551.0	Wet
3	1993-94	165.0	Dry
4	1994-95	285.0	Wet
5	1995-96	165.2	Dry
6	1996-97	117.0	Dry
7	1997-98	240.5	Wet
8	1998-99	232.2	Wet
9	1999-00	140.4	Dry
10	2000-01	144.0	Dry
11	2001-02	193.0	Dry
12	2002-03	291.5	Wet
13	2003-04	298.4	Wet
14	2004-05	343.0	Wet
15	2005-06	182.5	Dry
16	2006-07	272.5	Wet
17	2007-08	81.5	Dry
18	2008-09	151.5	Dry
Mean	-	232.09	-

*Please note that years with >200 mm are considered wet; those with <200 mm are considered dry.

Soil

The debris cone and the alluvial fan formed by the Bisheh Zard Ephemeral River is covered with a layer of drifting fine sand ranging in thickness from a few mm to several cm. A structureless, coarse sandy loam with average sand, silt and clay contents of 73.2, 14.5 and 12.2%, respectively, forms the A horizon, 10-20 cm thick. The stony C horizon lies directly under the A horizon (Kowsar, 1991). This soil has been classified as coarse-loamy skeletal, carbonatic (hyper) thermic, Typic Haplocalcids (Soil Survey Staff, 1999).

It should be noted that the deposition of nutritious suspended particles has changed the surface texture of the ARG systems to a fertile silt loam. Table 4, which reports soil texture modification after 5 years of spate irrigation, points to the gradual improvement of the soil quality in the ARG systems. Moreover, as the top-dressed dung and litter carried by floodwater in each event was covered by the next sedimentation, the organic matter of the soil has increased as well. These two positive changes have substantially enhanced the volumetric soil water capacity, and an increase in the soil depth has increased its available water capacity.

Table 2. Date and amount of rainfall on the Gareh Bygone Plain for the1991-2009 period.

Date			Precipitation, mm	Date			Precipitation, mm	Date			Precipitation, mm
Year	Month	Day		Year	Month	Day		Year	Month	Day	
1991	November	24	3	1992	March	2	1	1993	January	9	38
	December	2	7.5			18	16			10	3
		3	3.5			31	27			11	9
		5	15		April	24	1			12	10.5
		6	1.5		May	2	1			13	9
		7	34*			10	4.5		February	1	3.5
		8	6.5*			11	12			2	10
		16	2.5			12	4			3	39 *
		18	16.5		December	4	3			4	49 *
		19	1			9	26			5	6.5
		22	6.5 *			15	4.5			13	11.5
		25	1			17	3			14	10.5
1992	January	1	0.5			20	21.5			23	10.5
		2	6			21	56 *		March	4	13
		3	19 *			22	66 *			5	5
		12	7.5			23	1.5			6	2
		13	0.5			26	1.5			8	6
		17	17			27	4			17	1
		18	3			28	5.5			18	1
		21	31 *			29	11			21	0.5
		22	1			30	32.5		April	11	1
		31	9			31	1			12	4
	February	1	6	1993	January	2	5			27	4
		2	4.5			4	22		December	4	3.5
		10	2.5			5	38.5	1994	January	16	1.5
		25	1			6	1		January	24	4

*Flood-inducing rainfall

Table 2. (continued)

Date			Precipitation, mm	Date			Precipitation, mm	Date			Precipitation, mm
Year	Month	Day		Year	Month	Day		Year	Month	Day	
1994	January	25	3	1994	November	-	-	1995	September	14	2.5
	February	24	1.5		December	4	30.5 *	1996	February	20	2.5
		28	3			11	3.5			21	3.7
	March	1	5	1995	January	1	5			27	4.5
		8	9.5			25	1			28	8.5
		10	24 *			26	1.5		March	3	6
		11	27.5 *		February	5	8.5			4	23
		12	13			6	53 *			5	0.5
		16	6			7	24 *			7	1.5
		23	3.5			8	22 *			8	8
		26	24 *			9	4			10	20
		27	1			18	5			11	22
	April	1	2			23	5			21	2
		2	3		March	12	1.5			22	2.5
	May	1	2			13	3.5			23	9
		6	10.5 *			14	16.5 *			24	3.5
	August	7	4 *			20	3.5 *		April	15	2.5
	September	2	10 *			21	16.5 *			16	19
		4	3.5 *			22	5.5 *			17	13
	October	29	1.5		April	21	1		May	28	3
	November	8	3.5			22	1			29	3.5
		16	1.5			23	10 *		June	15	7
		17	14*			24	22 *		December	16	1.5
		18	1.5		May	30	4			26	3.5
		26	9 *		July	22	1	1997	January	15	0.5
		-	-			23	2.5			28	10

*Flood-inducing rainfall

Table 2. (continued)

Date			Precipitation,	Date			Precipitation,	Date			Precipitation,
Year	Month	Day	mm	Year	Month	Day	mm	Year	Month	Day	mm
1997	February	25	17	1998	January	13	11	1999	February	27	1.5
		26	1.5			26	3			28	5
	March	15	1			29	13		March	1	36 *
		16	1.5		February	12	30		July	8	6 *
		20	10			20	6.5		August	16	15 *
		21	30.5 *			22	5		December	12	1.8
		22	6.5 *		March	1	15	2000	January	7 to 10	16.9
		23	10 *			7	1.5			16	10.3
		27	1			8	4.5			17 to 19	68 *
		31	17.5			18	11			29	37 *
	April	4	1			19	26		February	28	2.7
		8	1			26	5			29	3.7
		9	3			27	6.5		October	4	6 *
	October	23	5		April	23	1.5		November	6	9
		31	4.5	1999	January	11	6 *			9	11 *
	November	11	5.5			15	1.2			18	4
		12	1.5			16	3			19	1
		27	6			19	27			20	0.5
	December	1	4.5			25	4		December	10	3
		4	1		February	4	4.5			11	33 *
		10	5			5	54*			15	19 *
		11	2			6	23*	2001	January	9	12 *
		14	5.5			13	5		February	11	8
		21	18			20	1.5			12	1
		30	0.5			21	36*			21	8
1998	January	2	27 *			22	1.5		March	17	13
		6	15 *			24	2		June	16	1.5

*Flood-inducing rainfall

Table 2. (continued)

Date			Precipitation,	Date			Precipitation,	Date			Precipitation,
Year	Month	Day	mm	Year	Month	Day	mm	Year	Month	Day	mm
2001	June	17	4	2002	March	28	20 *	2003	March	7	2
	August	11	3.5		April	12	8 *			13	1
	September	21	3.5		December	2	1.5			14	5
		22	3			10	0.5			20	7.5
	November	13	3			12	2			23	21 *
	December	5	1.5			13	4.5			26	22 *
		6	6.5			14	0.5			27	13
		14	7			18	0.5		April	14	3.5
		16	21 *			22	33 *			22	19 *
2002	January	2	2.5			25	1		July	23	32 *
		8	17 *	2003	January	1	6		August	4	2 *
		9	10		February	4	5			5	2 *
		10	12			16	0.5			17	3.5 *
		11	6			17	6		November	20	1.9
		12	30 *			23	8		December	2	13.5
	February	14	2.5			28	2.5			3	1.5
		15	3			29	14			5	11.5*
		20	12 *			5	25*			6	46*
		25	1			6	2*			7	8.5*
		26	2.5			14	0.5	2004	January	4	3.8
		27	6			15	0.5			7	1
		28	2.5			20	5.5			8	45 *
	March	18	11 *			21	19			9	60 *
		21	1			24	6			10	14
		22	7			25	14 *			14	33.5 *

*Flood-inducing rainfall

Table 2. (continued)

Date			Precipitation, mm	Date			Precipitation, mm	Date			Precipitation, mm
Year	Month	Day		Year	Month	Day		Year	Month	Day	
2004	January	15	3	2004	December	26	41 *	2005	November	21	2
		25	1	2005	January	7	2			22	1
		27	20.5 *			11	6		December	25	4.5
	February	3	4			18	6	2006	January	13	2.5
		6	1			19	29.5*			14	26.5
		7	1			20	31*		February	10	2.5
	March	19	8			23	30.5*			11	1
		20	6		February	13	20			13	3.5
	April	5	11			14	5			16	1.5
	August	27	1.2			22	3			22	1
	September	22	1.5			24	2			23	4.5
	November	24	1			25	2.5		March	29	8
		28	6.5			26	8.5			30	20*
	December	9	34 *		March	15	1			31	2.5
		10	11 *			16	6		April	6	2
		13	38 *			17	5			26	27
		14	17*		November	12	5			27	4
		15	2.5			13	18		July	12	1.5
		16	9			14	2			13	1.5
		17	19*			15	1		August	7	9.5
		25	6			17	17.5			8	1

*Flood-inducing rainfall

Table 2. (continued)

Date			Precipitation,	Date			Precipitation,	Date			Precipitation,
Year	Month	Day	mm	Year	Month	Day	mm	Year	Month	Day	mm
2006	September	5	11.5*	2007	February	5	1	2008	January	14	5.5
	November	16	3			14	9.5		April	9	2.5
		19	0.5			19	1			10	2.5
		26	9			20	0.5		May	9	0.5
	December	1	4		March	1	2			12	1
		2	6			15	0.5		September	8	18*
		4	1.5			16	0.5		November	15	2
		7	5.5			17	1	2009	January	2	8.5
		8	11.5			28	14.5*			3	0.5
		11	5			29	13.5*		March	1	29
		12	3		April	2	2			2	5
		17	37*			4	1			22	3.5
		18	11.5			5	3			26	4
		24	14*			12	0.5			30	54.5*
		25	1			13	2			31	3.5
2007	January	12	11			15	1.5		April	3	0.5
		17	7		July	24	0.5			4	1.5
		21	2			25	23*			10	8.5
		22	6.5		December	3	4			11	22.5*
		23	5.5			22	0.5			12	1
		31	1	2008	January	3	1.5			13	6.5
	February	1	1			4	30.5				
		2	5			5	4				
		3	3.5			11	7				
		4	40*			13	4				

*Flood-inducing rainfall

Table 3. Date, estimated peak flow, duration, and the volume of floodwater diverted from the Bisheh Zard Ephemeral River for the 1991-2009 period. Please note that the gauging station of this stream has been operative since December 2002.

No.	Date			Estimated peak flow, $m^3 s^{-1}$	Duration, hours	Volume of diversion, million m^3
	Year	Month	Day			
1	1991	January	26		14	
2		February	21-22		11	
3		Feb.-Mar.	27-1		40	
4		March	5-7		36	
5		March	13-14		29	
6		March	29-30		12	
7		August	8			
8		December	8			
9		December	22			
10	1992	January	3-4		16	
11		January	21-22		30	
12		December	21-22			
13	1993	February	3-4			
14	1994	March	10-11			
15		March	26			
16		May	6		5	
17		August	7			
18		August	15-16		14	
19		September	2&4			
20		November	18			
21		November	26		23	
22		December	4		23	
23	1995	February	6-8			
24		March	14,20,21&22			
25		April	23-24			
26		July	24-25		15	
27		December	6			
28		December	12			
29		December	25			
30	1996	January	4			
31	1997	March	21-23		11	
32	1998	January	2		4	
33		January	6			
34	1999	January	12			
35		February	5-6 & 21			
36		March	1-2			
37		July	8			
38		August	16-17			
39	2000	January	17-19&29			
40		October	4-5			

Table 3. (Continued)

No.	Date Year	Month	Day	Estimated peak flow, $m^3 s^{-1}$	Duration, hours	Volume of diversion, million m^3
41	2000	November	10			
42		December	12&15			
43	2001	January	10			
44		December	16			
45		January	8			
46		January	12			
47		February	21-22			
48	2002	March	18			
49		March	28			
50		April	12			
51		December	22	30	8	0.45
52		February	5-6	50	20	1.8
53		February	25	25	7	0.4
54		March	23	45	17	1.1
55		March	26	100	16	2.5
56	2003	April	22			
57		July	23	15	7	0.162
58		August	4-5	16.2	7	0.116
59		August	17	29	11	0.417
60		December	6	50	7	0.63
61		December	7	94	14.5	2.51
62		January	8	120	11.45	2.53
63		January	9	300	17.15	9.31
64		January	14	60	16.1	1.75
65	2004	January	27	30	6.45	0.364
66		December	9-10	50	14	1.26
67		December	13-14	215.5	14	5.43
68		December	17	250	17	7.65
69		December	26	290	15	7.83
70	2005	January	19-20	130	24	5.61
71		January	23	110	14	2.77
72		March	30	20	5.5	0.158
73	2006	September	5	30	3.3	0.178
74		December	17	60	18	1.994
75		December	24	20	4	0.144
76		February	4	120	10	1.26
77	2007	March	29	20	9	0.324
78		July	25	15	9	0.243
79	2008	September	9	30	4.5	0.1944
80		March	30	180	13.3	7.29
81	2009	April	11			

Miller et al. (1969), evaluated spate irrigation on more than 100 plots of rangeland in the western USA and concluded that for the production of 1,110-kg per ha of dry matter, the depth of water stored in the A and B_2 horizons should be equal to at least 12-cm at "field capacity". The highest yield, more than 4 tons per ha, was obtained on soils whose A and B_2 horizons' water content at "field capacity" ranged 20-40-cm. It should be emphasized that these rangelands receive summer storms that are highly effective in increasing forage yield.

Soil water characteristic curves presented in Fig. 2 clearly depict the gradual improvement in the water retention status of the sandy soil of the GBP.

Likewise, the dissolved nitrate and ammonium in the floodwater have increased the concentration of these two compounds in the soil (Tables 5 and 6). These, and the gradual increase in the organic phosphorus released through the decomposition of eucalyptus leaves (Khanmirazei et al., 2009) have prepared a better medium for plant growth than the original sandy soil.

Table 4. Effects of 5 years of suspended load deposition on selected physico- chemical properties of the 0-30 cm depth of a sandy soil in the Gareh Bygone Plain, Iran (after Naderi et al., 2000).

Parameters	Spate-irrigated	Control
Sand%	70.4*	73.2
Silt%	17.7	14.5
Clay%	11.9	12.2
Saturation percentage	25.0	23.4
Soluble K, mmol/l	0.28	0.25
Soluble Na, mmol/l	1.64	0.94
SolubleCa+Mg,mmol/l	3.95	3.95
Organic C, %	0.17	0.20
Total N, %	0.034	0.034
EC, ds/m	0.58	0.47
Available K, mg/kg	148	156
Available P, mg/kg	4	4
$CaCO_3$	38.11	38.28
pH (saturated paste)	7.98	7.96
CEC, cmol/kg	6.3	6.3
SAR	1.16	0.67
Exchangeable Na, cmol/kg	0.20	0.17
ESP	3.3*	2.8

* Significant at the 5% level.

Table 5. Nitrate-N concentration at three depths in sedimentation basins, recharge pond and control, mg/kg (after Yazdian and Kowsar, 2003).

Sampling site	Depth			
	0-25cm	25-75cm	75-125cm	Mean[a]
1st sedimentation basin	11.50	10.25	10.25	11.00 ab **
2nd sedimentation basin	15.20	7.52	11.25	11.32 a **
3rd sedimentation basin	12.48	8.95	11.48	10.97 ab **
4th sedimentation basin	8.08	7.55	7.55	7.72 b ns
5th sedimentation basin	15.83	9.12	7.77	10.91 ab **
Recharge pond	10.97	12.80	10.15	11.31 a **
Mean	12.35a	9.60b	9.85b	10.59 ab **
Control	6.93	7.15	6.61	6.88

[a]Means for the sedimentation basins and recharge pond followed by the same lower case letter or letters are not significantly different at the p=0.05 level, based on the Duncan's Multiple Range test. The difference between each sedimentation basin or recharge pond and the control is indicated by **, significant at the 1% level; ns: not significant.

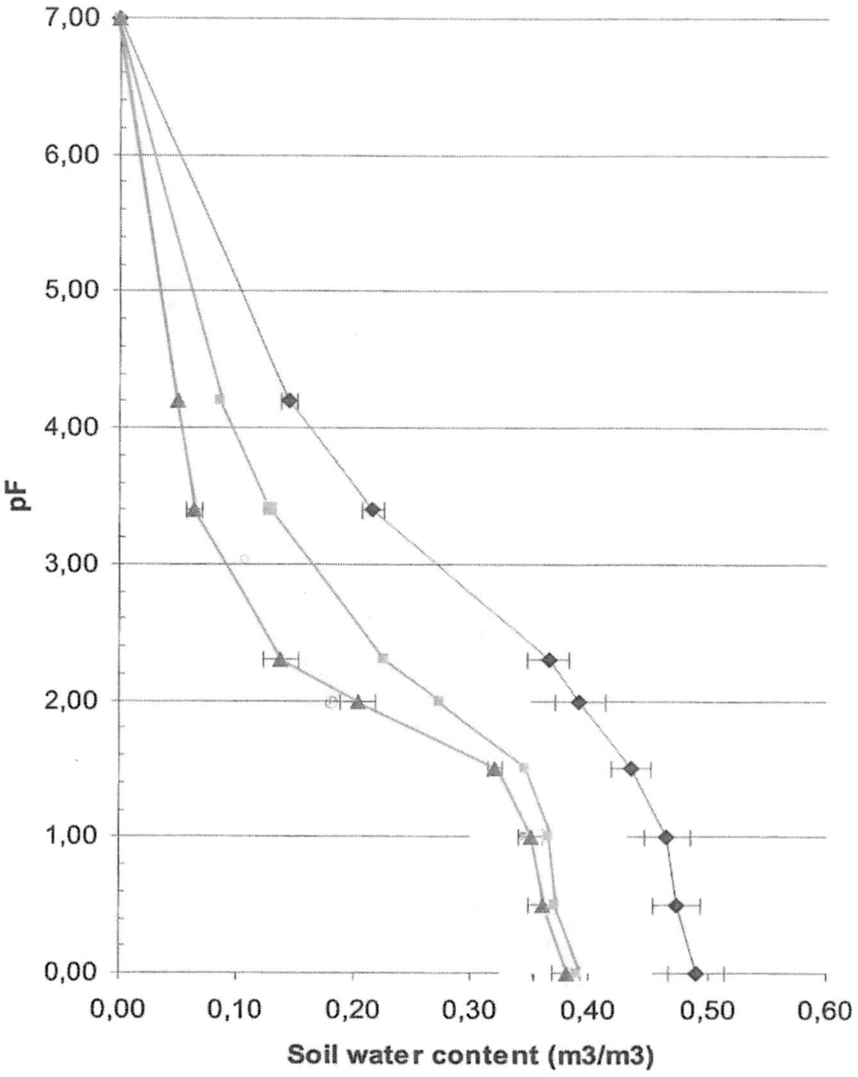

Figure 2. Water characteristics of the freshly- laid sediment (blue line), modified sand (orange line), and sand (green line). The error bars represent the standard error for each of the values (after Corens, 2005).

Vegetation

Phyto-geographically, the GBP is located between two major habitats, the Irano-Turanian domain and the Persian Gulf-Oman group. *Zizyphus nummularia* (Burm. F.) Wighth & Arn. and *Pteropyrum aucheri* Jaub& Spach, which attain a height of 3.0 and 1.5 m, respectively, are the tallest bushes on the debris cone. Occasional isolated bushed of *Amygdalus liciodes* Spach, *Amygdalus scoparia* Spach, and *Pistacia khinjuk* Stocks occupy hard to reach crevices; otherwise they were cut down (Kowsar, 1991). The list of the genera and species found in the basins and on the debris cones are given in Table 7; those that occupy the ARG systems and the control plots are presented in Table 8. The presence of a species in the spate-irrigated area indicates adaptation to a suitable growth medium; the lack indicates suppression, probably by sedimentation.

Table 6. Ammonium-N concentration at three depths in sedimentation basins, recharge pond and control, mg/kg (after Yazdian and Kowsar, 2003).

Sampling site	Depth			
	0-25cm	25-75cm	75-125cm	Mean
1st sedimentation basin	11.25	12.45	12.22	11.97 b ns
2nd sedimentation basin	15.90	12.22	18.08	15.42 ab *
3rd sedimentation basin	13.20	11.85	13.03	12.69 b ns
4th sedimentation basin	14.00	11.65	9.70	11.78 b ns
5th sedimentation basin	26.05	17.08	14.37	19.17 a **
Recharge pond	19.47	19.83	13.48	17.60 a **
Mean	16.65 a	14.18ab	13.48 b	14.77 ab *
Control	12.98	10.11	7.54	10.18

Means for the sedimentation basins and recharge pond followed by the same lower case letter or letters are not significantly different at the p=0.05 level, based on the Duncan's Multiple Range test. The difference between each sedimentation basin or recharge pond and the control is indicated by **, significant at the 1% level;* significant at the 5 % level; ns: not significant.

Spate Irrigation Systems for the Artificial Recharge of Groundwater

Ten ARG systems, covering 2445-ha in the GBP, annually capture about 10 million m^3 of floodwater on average of which approximately 80% reach the watertable. Each ARG system consists of an inundation canal that diverts floodwater to the command area. A stilling basin, called conveyor-spreading channel, 0.27-7.30 km in length, spreads the water as a thin, nonerosive sheet over the head of the first sedimentation basin. Most of the bed-load and coarse-grained particles settle in the canal, the stilling basin and the first sedimentation basin. The less turbid water enters the next sedimentation basin through gaps formed in the bank of a level-silled channel. This process is repeated 3-10 times depending on the flow rate and turbidity of floodwater. The mostly clear water is drained into a recharge pond at the end of the system. A spillway at the end of a tail drain returns the surcharge of the system into the streambed (Kowsar, 1991).

Forage Yield Assessment

Forage yield of the ARG systems were sporadically estimated from 1984 to 1991, when it was decided to do it on a more methodical and permanent basis. Therefore, in April 1991, ten, $1m^2$ temporary plots were established at random in 3 of the ARG systems, 5 parallel to the major slope of the land and 5 on the contour line. All of the forage species, disregarding their palatability, were determined in each plot and the percentage of area covered by grasses, forbs, shrubs and litter was determined using a $1m^2$ frame with a 10×10 grid. All of the grasses and forbs in the 30 plots were clipped at 1cm height and air-dried to attain a constant weight. The current year growth of shrubs were also sheared and dried as before.

Table 7. List of the genera and species found in the basins and on the debris cones in the Gareh Bygone Plain.

Number	Genra and species	Family	Number	Genra and species	Family
1	*Acacia sp.*L.	*Mimosaceae*	31	*Astragalus myriacanthous* Boiss.	*Papilionaceae*
2	*Acanthophyllum sp.*C.A.Mey.	*Caryophyllaceae*	32	*Astragalus onobrychis* L.	*Papilionaceae*
3	*Acantolimon sp.*Boiss.	*Plumbaginaceae*	33	*Astragalus podolobus* Boiss.&Hohen.	*Papilionaceae*
4	*Achillea eriophora* Dc.	*Asteracea*	34	*Astragalus sieberi* Dc.	*Papilionaceae*
5	*Aegilops cerasa* Boiss.	*Poaceae*	35	*Astragalus sp.*L.	*Papilionaceae*
6	*Aegilops triuncialis* L.	*Poaceae*	36	*Astragalus squarrosus* Burge.	*Papilionaceae*
7	*Ajuga chamaecistus* Giny.	*Lamiacae*	37	*Astragalus tribulioides* Delil.	*Papilionaceae*
8	*Alhagi camelorum* Fisch.	*Popilionaceae*	38	*Astragalus thalimansorensis* Sirj. & Rech.f.	*Papilionaceae*
9	*Allium sp.*L.	*Liliaceae*	39	*Atriplex lentiformis* S.Wats.	*Chenopodiaceae*
10	*Althea sp.*L.	*Malvaceae*	40	*Atriplex leucoclada* (Boiss.)Aellen	*Chenopodiaceae*
11	*Alyssum sp.*L.	*Brassicaceae*	41	*Avena fatua* L.	*Poaceae*
12	*Amygdalus lycioides* Spach	*Rosaceae*	42	*Biscutella didyma* L.	*Brassicaceae*
13	*Amygdalus scoparia* Spach	*Rosaceae*	43	*Bromus danthoniae* Trin.	*Poaceae*
14	*Amygdalus eburnean* Spach	*Rosaceae*	44	*Bromus sp.*L.	*Poaceae*
15	*Anchusa sp.*L.	*Boraginaceae*	45	*Calendula persica* C.A.Mey.	*Asteraceae*
16	*Andrachne sp.*L.	*?Ephedraceae*	46	*Capparis spinosa* L.	*Capparidaceae*
17	*Andrachne telephiodes* L.	*Euphorbiaceae*	47	*Carex stenophylla* L.	*Cyperaceae*
18	*Anisosciadium orientale* Dc.	*Apiaceae*	48	*Carthamus oxyacantha* M.B.	*Astraceae*
19	*Anthemis altissima* L.	*Asteracea*	49	*Cenchrus ciliaris* L.	*Poaceae*
20	*Anthemis microcephala* (Schrehk.)B.Fedtsch.	*Asteraceae*	50	*Centaurea bruguieriana* (Dc.)Hand.Mzt.	*Astraceae*
21	*Anvillea garcini* (Burm.)Dc.	*Asteraceae*	51	*Centaurea behen* L.	*Asteraceae*
22	*Aristida adscensionis* L.	*Poaceae*	52	*Chardinia orientalis* Desf.	*Asteraceae*
23	*Aristida caerulescens* Desf.	*Poaceae*	53	*Cichorium intybus* L.	*Asteracea*
24	*Artemisia sieberi* Besser.	*Astraceae*	54	*Cirsium sp.*Miller	*Asteraceae*
25	*Astragalus meridionalis* Bunge Sh.	*Papilionaceae*	55	*Citrullus colocynthis* (L.) Schrad.	*Cucurbitaceae*
26	*Astragalus curuirostris* Boiss.	*Papilionaceae*	56	*Cleom oxypetala* Bioss.	*Capparidaceae*
27	*Astragalus cephalanthus* Dc.	*Papilionaceae*	57	*Convolvulus acathocladus* Bioss.	*Convolvulaceae*
28	*Astragalus eburneus* Bornm.& Gauda.	*Papilionaceae*	58	*Convolvulus buschiricus* Bornm.	*Convolvulaceae*
29	*Astragalus fasiculifolius* Bornm. & Gauba.	*Papilionaceae*	59	*Convolvulus leiocalycinus* Boiss.	*Convolvulaceae*
30	*Astragalus hamadanus* Boiss.	*Papilionaceae*	60	*Convolvulus sp.*L.	*Convolvulaceae*

Table 7. (continued)

Number	Genra and species	Family	Number	Genra and species	Family
61	*Cornulaca monacantha* Delil.	*Chenopodiaceae*	91	*Isatis minima* Bye.	*Brassicaceae*
62	*Crupina crupinastrum* (Moris)Vis.	*Asteracea*	92	*Juncus sp.*L.	*Juncaceae*
63	*Cymbopogon olivieri* (Boiss.)Bor.	*Poaceae*	93	*Launea acanthodes* (Boiss.) O.Kuntze	*Asteraceae*
64	*Cynodon dactylon* (L.)Pers.	*Poaceae*	94	*Lolium sp.*L.	*Poaceae*
65	*Dendrostellera lessertii* (Wikstr.)VanTiyeh.	*Thymelaeaceae*	95	*Lophochloa sp.*Reichenb.	*Geramineae*
66	*Dianthus sp.*L.	*Caryophyllaceae*	96	*Lycium schawii* Roemer and Schult.	*Solanaceae*
67	*Ducrosia anethifolia* (Dc.) Boiss.	*Apiaceae*	97	*Malva sylvestris* L.	*Malvaceae*
68	*Ebenus stellata* Boiss.	*Popilionaceae*	98	*Marrubium cuneatum* Russell	*Lamioaceae*
69	*Echinops endotrichus* Rech.F.	*Asteracea*	99	*Matthiola sp.*R.Br.	*Brassicacea*
70	*Echinops sp.*L.	*Asteracea*	100	*Medicago polymorpha* L.	*Papilionaceae*
71	*Ephedra foliate* Boiss. and Kotschy	*Ephedraceae*	101	*Medicago radiata* L.	*Papilionaceae*
72	*Eremostachys laevigata* Bunye.	*Lamiacae*	102	*Nerium oleander* L.	*Apocynaceae*
73	*Eremostachys sp.*Bye.	*Lamiacae*	103	*Noaea mucronata* (Forsk.)Aschers.et Schweivf.	*Chenopodiaceae*
74	*Erucaria sp.*Gaertn.	*Brassicaceae*	104	*Ochradenus baccatus* Delil.	*Resedaceae*
75	*Eucalyptus camaldulensis* Dehnh.	*Myrtaceae*	105	*Onobrychis aucheri* Boiss.	*Papilionaceae*
76	*Euphorbia sp.*L.	*Euphorbiaceae*	106	*Onobrychis crista – galli* (L.)Lam.	*Papilionaceae*
77	*Fagonia acerosa* Boiss.	*Zygophyhllaceae*	107	*Ononis serrata* Forssk.	*Papilionaceae*
78	*Ferula asafetida* L.	*Apiaceae*	108	*Onopordon sp.*L.	*Astraceae*
79	*Ferula sp.*L.	*Apiaceae*	109	*Onosma sp.*L.	*Boraginaceae*
80	*Ferulago sp.* Koch	*Apiaceae*	110	*Orobanche sp.*L.	*Orobanchaceae*
81	*Ficus sp.*L.	*Moraceae*	111	*Otostegia persica* (Burm.)Boiss.	*Lamiacae*
82	*Gagea sp.*Salisb.	*Liliaceae*	112	*Papaver sp.*L.	*Papaveraceae*
83	*Galium aparine* L.	*Rubiaceae*	113	*Peganum harmala* L.	*Zygophyhllaceae*
84	*Glycyrrhiza glabra* L.	*Papilionaceae*	114	*Pennisetum sp.*Rich.	*Poaceae*
85	*Gundelia tournefortii* L.	*Astraceae*	115	*Phalaris minor* Retz.	*Poaceae*
86	*Gymnocarpus decander* Forssk.	*Caryophyllaceae*	116	*Phlomis sp.*L.	*Lamiacae*
87	*Hammada salicornica* (Moq.) Iljin	*Chenopodiaceae*	117	*Planthyos coronopus* L.	*Plantaginaceae?*
88	*Haplophyllum sp.*A.Juss.	*Rutaceae*	118	*Picris strigosa* M.B.	*Asteracea*
89	*Helianthemum lippii* (L.)Pers.	*Cistaceae*	119	*Pimpinella sp.*L.	*Apiacea*
90	*Hordeum marinum* Hudson	*Poaceae*	120	*Platychaeat aucheri* (Boiss.)Boiss.	*Asteraceae*

Table 7. (Continued)

Number	Genra and species	Family	Number	Genra and species	Family
121	*Plntago lagopus* L.	*Plantaginaceae*	151	*Urginea sp.*Steinh.	*Liliaceae*
122	*Poa bulbosa* L.	*Poaceace*	152	*Urtica sp.*L.	*Urticaceae*
123	*Popolus euphratica* Olivier	*Salicaceae*	153	*Verbascum farsistanicum* (Murb.) Hub.Mor.	*Scrophulariaceae*
124	*Prosopis farcta* (Banks and Soland.)Machr	*Mimosaceae*	154	*Vicia sp.*L.	*Papilionaceae*
125	*Pteropyrum aucheri* Jaub and spach	*Polygonaceae*	155	*Vitex pseudo- Negundo* (Hausskn.) Hand- Mzt.	*Verbenaceae*
126	*Pycnocylca nodiflora* Decne ex Boiss.	*Apiaceae*	156	*Zataria multiflora* Boiss.	*Lamiacae*
127	*Reseda luteola* L.	*Resedaceae*	157	*Ziziphora teniur* L.	*Lamiacae*
128	*Rheum sp.*L.	*Polygonaceae*	158	*Ziziphus nummularia* (Burm.F.)Wiyhth and Arn.	*Rhamnaceae*
129	*Salsola sp.*L.	*Chenopodiaceae*	159	*Zygophyllum atriplicoides* Fisch. & C.A.Mey.	*Zygophyhllaceae*
130	*Salvia eremophila* Boiss.	*Lamiacae*			
131	*Salvia sp.*L.	*Lamiacae*			
132	*Scabiosa olivieri* Coult.	*Dipsacaceae*			
133	*Scandix sp.*L.	*Apiaceae*			
134	*Scariola orientalis* (Boiss.)Sojak	*Asteraceae*			
135	*Scrophularia sp.*L.	*Scrophulariaceae*			
136	*Stellaria media* (L.)Cyr.	*Caryophyllaceae*			
137	*Stipa Arabica* Desf.	*Poaceae*			
138	*Stipa capensis* Thunb.	*Poaceae*			
139	*Stipagrostis plumose* (L.)Munro ex T.Anders	*Poaceae*			
140	*Suaeda sp.*Forssk.	*Chenopodiaceae*			
141	*Tamarix aphylla* (L.)Krasten	*Tamaricaceae*			
142	*Tecomella undulata* (Roxb.)Seem.	*Bignoniaceae*			
143	*Teucrium polium* L.	*Lamiacae*			
144	*Tragopogon sp.*L.	*Asteraceae*			
145	*Trigonella spruneriana* Boiss.	*Papilionaceae*			
146	*Trigonella elliptica* Boiss.	*Papilionaceae*			
147	*Trigonella spp.*L.	*Papilionaceae*			
148	*Tulipa sp.*L.	*Liliaceae*			
149	*Turgenia latifolia* (L.)Hoffm.	*Geraniaceae*			
150	*Typha australis* Schum.	*Typhaceae*			

Table 8. List of genera and species found in the research plots.

No	Name	Control	Spate-irrigated
1	*Acantolimon sp.*Boiss.	*	*
2	*Achillea eriphora* Dc.	-	*
3	*Aegilops cerasa* Boiss.	*	*
4	*Aegilops triuncialis* L.	-	*
5	*Alhagi camelorum* Fisch.	*	*
6	*Allium sp.* L.	-	*
7	*Anisosciadium orientale* Dc.	*	*
8	*Artemisia sieberi* Besser.	-	*
9	*Astragalus onobrychis* L.	-	*
10	*Astragalus squarrosus* Burge.	*	-
11	*Astragalus thalimansorensis* Sirj. & Rech.f.	*	*
12	*Atriplex leucoclada* Boiss.	-	*
13	*Avena fatua* L.	*	*
14	*Bromus danthoniae* Trin.	-	*
15	*Bromus sp.* L.	-	*
16	*Carex stenophylla* L.	*	*
17	*Carthamus oxyacantha* M.B.	-	*
18	*Centaurea bruguieriana* (Dc.)Hand-Mzt.	*	*
19	*Cymbopogon olivieri* (Boiss.) Bor.	-	*
20	*Convolvulus sp.*L.	-	*
21	*Cynodon dactylon* (L.)Pers.	*	*
22	*Dendrostellera lessertii* (Wikstr.)Van Tiyeh.	*	*
23	*Echinops endotrichus* Rech.F.	*	*
24	*Ephedra foliate* Boiss. & Ky.	*	*
25	*Gundelia tournefortii* L.	-	*
26	*Haplophyllum sp.*A.Juss.	-	*
27	*Helianthemum lippii* (L.)Pers.	*	*
28	*Hordeum marinum* Hudson	*	*
29	*Lolium sp.*L.	-	*
30	*Medicago polymorpha* L.	*	*
31	*Medicago radiata* L.	-	*
32	*Noaea mucronata* (Forsk.)Aschers.et Schweinf.	*	*
33	*Peganum harmala* L.	*	*
34	*Phalaris minor* Retz.	-	*
35	*Pimpinella sp.*L.	*	*
36	*Poa bulbosa* L.	-	*
37	*Prosopis farcta* (Banks & Soland.) Machr	-	*
38	*Salsola sp.*L.	*	-
39	*Scariola orientalis* (Boiss.) Sojak	*	*
40	*Stipa capensis* Thunb.	*	*
41	*Stipagrostis plumose* (L.)Munro ex T.Anders	*	*
42	*Teucrium polium* L.	-	*
43	*Trigonella spruneriana* Boiss.	*	*
44	*Turgenia latifolia* (L.)Hoffm.	*	*
45	*Vicia sp.*L.	-	*
46	*Ziziphora teniur* L.	-	*

* present; - absent.

Statistical Analyses

Single-factor ANOVAs were performed using PROC GLM in SAS to examine treatment difference for the spate-irrigated and control plots combined over 18 years (SAS Institute, 2006). If a significant F test (F<0.05) was obtained from an ANOVA, Duncan's New Multiple Range test was used for determining treatment differences at P=0.05 . As the average yield and crown cover are of importance for our purpose, only these 2 data will be reported here. The same procedure was fallowed outside of the ARG system as the control.

Land Use Change Assessment

The earliest document that we could find of the project area was a Landsat-4, a 7-band imagery dated 16 July 1984, 19 months after initiation of the project. It should be noted that the first high yielding well that hit the watertable at a depth of 19 m in early June 1983 started a water rush in the GBP; therefore, the irrigated area shown in the 1984 imagery had certainly increased relative to that of Jan. 1983, when we started the ARG activities. However, we have no other alternative but to accept this imagery as the baseline. A Landsat5, a 7-band imagery of 20 May 1998, and an ASTER 3-band imagery of 26 July 2002 were also used in this study.

The required corrections were made on these imageries and all of them were digitized to facilitate data collection. The most important criterion employed for delineating the plain was an 8% slope of the foothills. Different land use of the study area was polygonized employing the visual interpretation method by digitizing them on the screen. The outputs were the area and the percentage of each of the land use types. These data were then statistically analyzed using the integrated land and water information system (ILWIS ver-3.1).

RESULTS

In this chapter, we report only on the results obtained in the Bisheh $Zard_1$ ARG system, as we have to substantiate our findings benefiting from the already published data on the improvement of the texture and nutrient status of the soil.

Spate irrigation has transfigured a desolate sandy desert into a relatively productive growth medium in which adapted trees, bushes and shrubs thrive remarkably and the indigenous plants grow abundantly. Recharging an over-exploited aquifer has encouraged reverse migration of the former sedentarized nomads to the GBP and the revitalization of the local economy. Plates 1– 4 present the state of the spate – irrigated rangeland 15 years after the initiation of range improvement project. Plates 1-4 depict how the nutritious floodwater can transform a desolate sandy desert into a productive pasture.

Plate 1. Spate irrigation of an improved rangeland. A quail bush block is seen in the middle of the picture.

Plate2. Spate irrigation has greatly increased the yield of indigenous range plants growing in a sandy soil.

Plate3. Nomadic herds graze and brows the indigenous and introduced range plants and fodder trees.

Plate4. *Hordeum marinum* Hudson and *Phalaris minor* Retz. have invaded the inundation canal of the Bisheh $Zard_4$ artificial recharge of groundwater system.

Forage Production

1. Indigenous Species

The annual and daily rainfall, the flooding events that resulted in spate irrigation, the frequency and mean annual yield of plant types, forage yield and crown cover for the 1992-2009 period are reported in Tables 1, 2, 3, 9, and 10, respectively. It is observed that there is a poor relationship between the amount of annual rainfall and yield (regression equations are not shown). A very important reason for this apparent anomaly is flooding of the basin is more critical than the depth of rain received in each event. For instance, the highest annual yield occurred in 1996, when the area received only 165.2 mm of precipitation, while we received a generous flooding in July and 3 flooding in December 1995, and one in January 1996.

For an event to be effective it should rain more than 30 mm to moisten the surface 30 cm of sand, and more than this amount for the freshly-laid, fine-grained sediments in the basins. However, we receive a nominal flood flow if 5 mm of rain with an intensity of 5 mm per hour occurs. Comparison of the yield of control plots with the annual rainfall supports this argument (Table10).

Table 9. Frequency and mean annual yield of plants in the spate-irrigated and control plots.

Parameters	Bisheh $Zard_1$		Control	
	Quantity	%	Quantity	%
Frequency of plants in a plot	46	100	26	100
Frequency of forage plants in a plot	35	76.09	8	30.77
Frequency of non - forage plants in a plot	11	23.91	18	69.23
Frequency of annual grasses in a plot	26	74.29	5	62.50
Frequency of perennial grasses in a plot	9	25.71	3	37.50
Mean yield of a plot, kg/ha	411.11	100	149.18	100.00
Mean yield of shrubs in a plot, kg/ha	97.34	23.68	5.07	3.40
Mean yield of grasses in a plot, kg/ha	198.56	48.3	54.10	36.26
Mean yield of forbs in a plot, kg/ha	115.21	28.02	90.01	60.34
Mean yield of annuals in a plot, kg/ha	313.77	76.32	144.11	96.60
Mean yield of perennials in a plot, kg/ha	97.34	23.68	5.07	3.40

Not having a better criterion to distinguish between dry and wet years, we arbitrarily set the borderline at 200 mm rainfall per year. Assuming that the allowable use coefficient (AUC) for our conditions is 50%, the usable forage yield during the 18-year research period ranged 136.27-721.50 kg $ha^{-1}yr^{-1}$, with the mean of 411.11 kg $ha^{-1}yr^{-1}$ for the spate-irrigated range plants. This, for the control was 78.00-350.20 kg $ha^{-1}yr^{-1}$, with the mean of 149.18 kg $ha^{-1}yr^{-1}$. The usable forage yield was 490.77 and 182.58 kg ha^{-1} yr^{-1} for the spate-irrigated and control, respectively, in the wet years; this was 331.45 and 115.82 kg ha^{-1} yr^{-1} for the spate-irrigated and control, respectively, in the dry years; thus, we had an extra production of 215.63 kg ha

Table 10. Mean annual yields and crown cover of the spate-irrigated and control plots for the1991-2009 period.

No.	Year	Condition	Yield, kg/ha		Crown cover, %	
			Spate-irrigated	Control	Spate-irrigated	Control
1	1992	Wet	655.00	85.60	49	20.1
2	1993	Wet	437.60	88.30	38	27.4
3	1994	Dry	317.80	87.00	37.8	15.7
4	1995	Wet	432.20	90.00	35.2	20.3
5	1996	Dry	721.50	125.60	36.4	18.3
6	1997	Dry	581.00	105.00	32.1	14.6
7	1998	Wet	457.00	98.00	23.6	12.6
8	1999	Wet	330.70	83.00	20.3	11
9	2000	Dry	257.30	78.00	19	9.7
10	2001	Dry	260.00	80.00	19.5	12.15
11	2002	Dry	136.27	90.10	26.98	20
12	2003	Wet	626.41	350.20	67.09	28
13	2004	Wet	563.70	313.70	30.32	15.65
14	2005	Wet	501.50	340.70	36.65	30.2
15	2006	Dry	325.01	190.50	31.03	22.79
16	2007	Wet	412.80	193.30	38.04	23.9
17	2008	Dry	180.24	145.80	18.59	24.38
18	2009	Dry	203.93	140.4	21.3	19.5
	Mean		411.11±172.62	149.18 ± 92.63	32.27 ± 12.2	19.24 ± 6.08

Table 11. Eighteen – year mean yield and crown cover of the spate- irrigated and control plots in "wet" and "dry" years.

Site	Wet		Dry	
	Yield, kg/ha	Crown cover, %	Yield, kg/ha	Crown cover, %
Bisheh $Zard_1$	490.77	37.58	331.45	26.97
Control	182.53	21.02	115.82	17.46
Significance, %	1	5	1	5

$^{-1}yr^{-1}$ during a typical drought. This amounts to a 186 % increase relative to the control.

Keeping in mind that we are mostly concerned with droughty years, the 215.63 ha^{-1} increase in allowable grazing in a typical drought on 2,070 ha of the treated area, which receive a fairly good watering, can support an additional 815 heads of livestock for a full year, assuming the daily intake of a 30 kg sheep or goat is 1.5 kg of forage. Moreover, the straw and stubble of 731 ha irrigated wheat and barley fields will feed another 6675 heads of sheep and goats for the same duration. Please note that the area of irrigated fields decreased from 1,193 to 731 ha due to water limitation during a severe drought.

Table 12. Number of livestock that grazed the spate-irrigated rangeland, grazing duration, and weight of the removed forage for years in which reliable data are available. F: E signifies the relative amount of forage removed from the rangeland.

Year	Number of livestock	Grazing duration, days	Forage removed(F), kg	Estimated yield(E), kg	F:E,%
1996	2800	31	130,200	1493505	8.7
1998	2487	31	115,645.50	945990	12.2
1999	1840	31	85,560	684549	12.4
2002	2303	30	103,635	282078	36.6
2003	3406	29	148,161	1296668	11.4
2004	2405	31	111,832.50	1166859	9.5
2005	2722	35	142,905	1038105	13.7
2007	1850	44	122,100	854496	14.2
2009	3700	45	249,750	422135	59.1

As the vegetative cover is vital for protecting soil surface against wind and water erosion, and soil building is an integral function of spate irrigation, therefore, we believe that the 50% AUC for this situation is too low. However, as assigning a higher AUC for the spate-irrigated rangeland may be considered a bias against the control, we reluctantly accepted this figure! To substantiate this claim, we report the number of livestock and the duration of grazing for the years we have kept reliable data (Table 12).

The grazing season for the spate-irrigated rangeland in the GBP is fall. Forage availability on free range during spring, and abundance of straw and stubble during summer usually keep the herds out of the ARG systems until fall. Although herbaceous species lose some of their nutritional value, much of their succulence, and decrease indigestibility due to drying, as the strong winds that blow in spring and summer carry the straw, the herders put the livestock on the fields as soon as the combine harvesters finish their work.

Three ARG systems, totaling 500 ha in area, supported 3,637 heads of sheep and goats for a period of 45 days in November and December 1987 (Kowsar, 1991). At 1.5 kg per head per day, the herds had removed 490 kg ha^{-1}, almost identical with our reported estimate of 490.77 kg ha^{-1} for an average wet year.

A controversial finding of this study is the preference of annuals over perennials :

"The establishment of perennials is desirable, for the annual grasses are of low feed value during the dry summer and fall periods, whereas perennials are able to remain green for a longer period and to maintain their nutritive value at a higher level when dry" (Stoddart and Smith, 1955).

"Annual plants depend upon a favorable period each spring during which they can germinate and send their roots to the moist subsoils. Such a period is far from a surety over most ranges, and, hence, failures are common. Fluctuation in forage volume from year to year is much greater on ranges high in annual plants. ...Perennials, having deep roots already established, are less dependent upon current precipitation and more upon temperature for their start of growth" (Stoddart et al., 1975)

"Seeding annual cereal crops on range land, especially for early-spring feed, has been recommended widely in recent years.Though certain conditions justify the use of such species, perennials generally are preferred"(Stoddart and Smith, 1955).

Establishment of perennials requires protection of seedlings from grazing for two to three growing seasons to develop a good rooting system to carry them during droughts, when they can benefit from the stored water in deeper horizons (Stoddart et al., 1975). It is practically impossible to keep the livestock out of green pastures in our deserts. Were it not for the wolves that attacked the nightly trespassers we would not have a reliable set of data to present in this chapter! Furthermore, as the annuals form a large percentage of the palatable portion of the forage and the perennials are mostly less palatable brush, therefore, we recommend seeding of adaptable annual grasses and legumes. This is in contrast with the long-established worth of perennials advocated by range ecologists.

The increase in forage production in a typical year for 2,070 ha of flooded basins is sufficient to support 864 heads of sheep and goats the year round. If this is added to the stubble of irrigated small grains then the nomads can keep about 8,462 livestock in the plain, assuming they do not graze the poor and very poor rangeland, and use the straw for maintenance of livestock as winter supplement.

The second observed anomaly is the reduction in crown cover as the years go by. We speculate that replacement of forage plants (elimination of some and introduction of a few) has been responsible for this unexpected observation. A research project to resolve this matter has been started.

2. Quail Bush (*Atriplex lentiformis* [Torr.] S. Wats.)

A 70% survival rate of quail bush seedlings planted in the first sedimentation basin of the Bisheh $Zard_1$ ARG system in March 1983encouraged us in planting this C_4 fodder on a rather large scale. Therefore, we planted the seedlings in the close-to-contour furrows spaced at 4 m intervals (625 bushes per ha) in Mar. 1984 in the Bisheh $Zard_1$and Bisheh $Zard_2$ systems. This was repeated in Jan.-Mar.1985in the Bisheh $Zard_3$, Bisheh $Zard_4$, Rahim $Abad_1$, and Rahim $Abad_2$ systems. These bushes, which cover about 150 ha, attained the height and crown diameter of 1.5-2.5 m in 3 years. The interlocking crowns of the bushes in many places have provided a safe refuge for wild boars and wolves. As we have not performed a methodical study on the yield of this species we can not offer reliable data on its annual harvest by the browsers. However, knowing the number of livestock, which brows the pasture and the length of the grazing season helps us in making a conservative estimate of 1ton ha^{-1} yr^{-1}.

Quail bush is the favorite forage of a sowbug (*Hemilepistus shirazi* Shuttz), which appeared in large numbers in 1992. This crustacean offers the most valuable service to our ARG systems by burrowing in the thick crust deposited by many years of turbid floodwater to form its nest. Infiltration rate of the areas containing the burrows is about 5 times that of control. The burrows, which have been measured to a depth of 1.8 m, facilitate preferential flow of water to the subsoil, thus the recharge of groundwater (Rahbar and Kowsar, 1997). Moreover, enhancement of the fertility status of the area surrounding the sowbug burrows by the excavated material heralds the initiation of a new life cycle in the GBP (Rahbar et al., 2001; Kowsar, 2007; Rahbar et al., 2007).

Alarming information about this useful fodder shrub prevented us from planting it on a much larger scale. As quail bush is an alternate host for the sugar beet curly top virus (Young et al., 1980), and a selenium accumulator (Davis, 1972), its introduction into sugar beet growing areas such as the GBP should be after consultation with plant pathologists and animal nutritionists. A preliminary study by Mr. Hans Paul Velema of the Wageningen University (personal communication) in the GBP did not show any evidence that quail bush was an alternate host of leafhopper vector (*Circulifer haematoceps* [Mulsant and Ray]), which transmits the curly top virus to the sugar beet plants. However, the abundance of this insect and the beet leafhopper (*Circulifer tenellus* [Baker]) in the area justifies a detailed study.

Quail bush can not tolerate inundation; those planted upstream of embankment that hold water for a week have died.

3. Fodder Trees

Inclusion of fodder trees in shelterbelts and windbreaks not only provide wildlife refuge and shade for the ruminating livestock, but also some fodder during droughts when these long-rooted plants can benefit from the stored water in deep soils. This allows the herders to keep a limited number of healthy animals as "seed" for the coming wet years.

We planted a limited number of elegant wattle (*Acacia victoriae* Benth.), willow wattle (*Acacia salicina* Lindl.), and orange wattle (*Acacia saligna* H.L. Wendl., Syn. *Acacia cyanophylla* Lindl.) in March 1983 in an elimination trial. As 63% of orange wattles succumbed to -7.0°C, we planted the other two species on a large scale in Jan. 1986. The organic matter content, and the concentration of total N, soluble K, Ca, Mg and EDTA extractable Mn were higher in the under canopy soil profile of these trees as compared with those of the between canopies (Kowsar, 2007, based on the M.S. thesis of Mr. Javad Fallah Shojaei).

Livestock readily consume the leaves, twigs, pods and seeds of these leguminous trees and honeybees frequent them profusely. Willow wattle is a very generous tree as it is the only plant in the area, which is in flower from Sept. to March. Honeybees visit this tree when air temperature exceeds 0°C.

A strange observation is worth mentioning here: sheep and goats severely browse leaves and twigs of certain *Eucalyptus camaldulensis* Dehnh. trees with no apparent ill effects (Kowsar, 1991). Whether they eat this apparently unpalatable fodder out of hunger or preference is not known. The glad tidings for the drought-stricken herders are the honey production potential of close to 100 ha of eucalyptus and acacia plantations. The reported annual honey yield of our station was 35 tons in 2006 (Kowsar, 2007); however, the potential is higher than this figure. Moreover, according to an unconfirmed report, the quality and quantity of honey are enhanced during droughts.

Irrigated Fields

Irrigated farm fields covered 147 ha in 1984. In 1998, they occupied 1193 ha, an 8-fold increase. This is the most remarkable improvement in the livelihoods of the nomads-turned farmers. Bakhtiar et al. (1997) have reported that 5.5 ha of irrigated fields can support a 6.7 member owner-operator family, and create 0.38 extra occupations for farm hands. Based on this criterion, the newly irrigated farm fields have created 262 well-paid occupations in the

GBP. Moreover, the stubble of these fields that exceed 5000 metric tons in a good year is a valuable food reserve for the livestock. A very encouraging observation in the wheat fields of the GBP is their relatively high yields (up to 5 ton $ha^{-1}yr^{-1}$) in loamy sand (> 73% sand) irrigated at a 3- week interval. This observation that requires a thorough study points to the potential of sandy deserts for food production.

LAND USE CHANGES

Figure 3 depicts the area of 9 categories of land use for the years 1984, 1998, and 2002; details are presented in Table 12. A very prominent change of bare land from 3,137 ha in 1984 to 1,067 ha in 1998 was mainly due to construction of the ARG systems. These systems, which covered 475 ha in 1984, occupied 2,445 ha in 1998. Please note that 375 ha of the ARG systems (15.3%) were still devoid of a good vegetative cover in 1998, because they were not irrigated properly. As the basins are leveled due to sedimentation, this bare land will also be covered with floodwater too. An increase of 1,504 ha in bare land area in 2002 was due to a drought during the 1997-2000 period. It is noteworthy that the ARG activities not only decreased the total area of bare land, but also provided forage for the livestock and reduced the grazing pressure on the rangeland. No new ARG system has been constructed since 1998.

The area of irrigated farm fields increased from 147 ha in 1984 to 1,193 ha in 1998; however, the drought of 1997 and the elongated dry period of 1999-2001 (Table 13), and a substantial decrease in natural and artificial recharge of groundwater, reduced the irrigated area to 731 ha in 2002. This is also evident in the area of the fallowed fields that increased by 211 ha in 2002 relative to 1998.

Table 13. Land use change during the 1984-2002 period (after Kowsar and Pakparvar, 2004).

Land use type	Year					
	1984		1998		2002	
	ha	%	ha	%	ha	%
Bare land	3,137	21.13	1,067	7.19	2,571	17.32
Irrigated fields	147	0.99	1,193	8.04	731	4.92
Rainfed fields	335	2.26	831	5.6	158	1.06
Fallowed fields	304	2.05	182	1.23	393	2.65
Spate-irrigated	475	3.2	2,445	16.47	2,445	16.47
Afforested	0	0	87	0.58	87	0.58
Poor rangeland	217	1.46	1,972	13.29	2,917	19.65
Very poor rangeland	10,224	68.89	7,047	47.48	5,497	37.04
Urban	3	0.02	18	0.12	44	0.3
Total	14,842	100	14,842	100	14,842	100

An alarming observation is a significant increase in the poor rangeland area: 217 ha in 1,984 vs. 2,917 ha in 2002. This is an indication of overgrazing that is happening in most drylands of the world. A positive sign, which is the direct result of the ARG activities, is a drastic decrease in the area of the very poor rangeland: 10,224 ha in 1984 vs. 5,497 ha in 2002. Assuming that 1970 ha of this area had been converted to the ARG systems, there still had been a reduction of 3,527 ha in the very poor rangeland expanse. Taking these 2 categories together, there has been a reduction of 827 ha in the area of poor and very poor rangeland in the GBP. As the sedentarized nomads have become mixed farmers, and livestock husbandry is still an inseparable part of their lives, provision of forage for sheep and goats cannot be overemphasized.

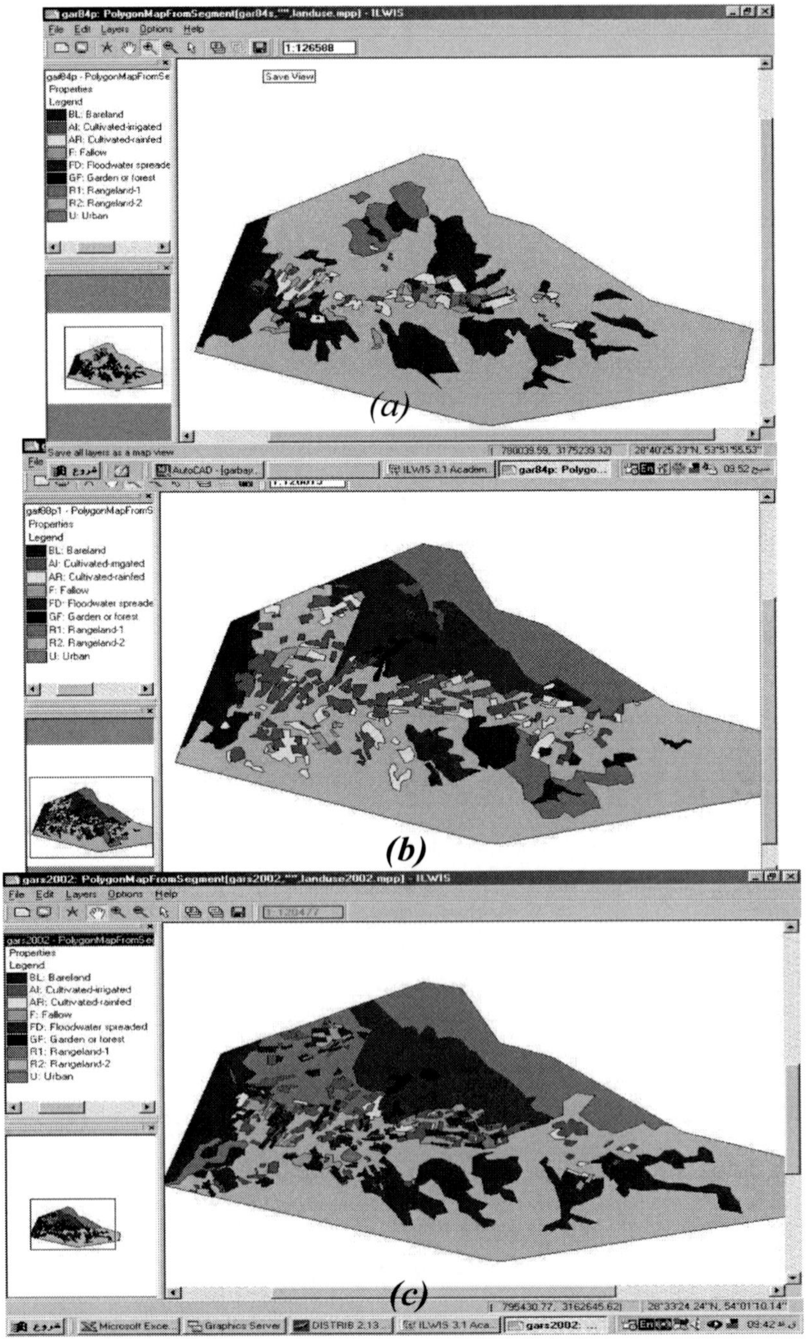

Figure 3. Land use changes during the 1984- 2002 priod.a.1984,b.1998,c.2002.

Spate-irrigated eucalyptus and acacia plantations cover 87 ha of the ARG systems. Furthermore, 30 km of windbreaks cover another 13 ha. Carbon sequestration, an important global issue deserves another publication. Suffice it to say that these trees annually sequester about 290 tons of carbon above ground.

An observation not yet proved by rigorous analysis supports the function of the freshly-laid, fine sediments in enhancing the available water capacity of the sandy soil. Comparison of the 18-year mean yield (Table 14) and sediment depth in Fig.4 reveals that the higher yield of the first sedimentation basin may be due to its larger sediment thickness, as the turbid floodwater loses most of its silt content in that basin. As the floodwater contains 17.4 and 13.4 milligrams of dissolved NH_4-N and NO_3-N per liter, respectively, and the suspended load also contains 4.3 and 1.3 milligrams of NH_4-N and NO_3-N per kilogram, respectively (Yazdian and Kowsar, 2003), it is logical to infer a higher fertility, thus productivity, for the 1st sedimentation basin relative to the rest of the ARG system as well.

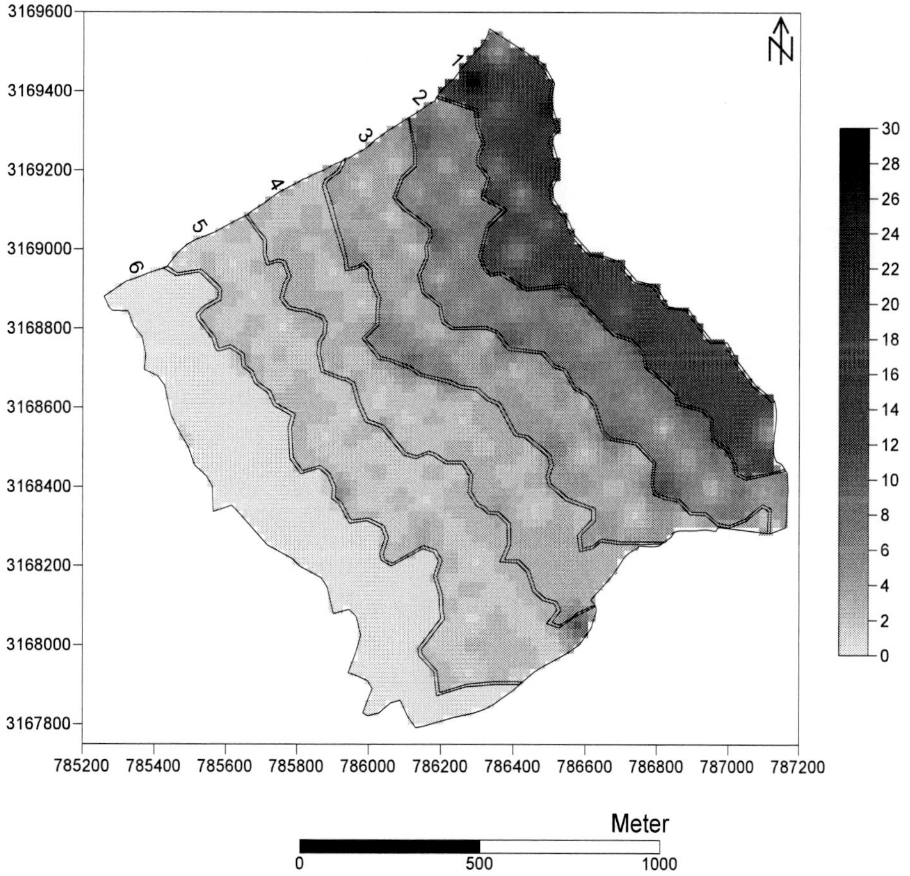

Figure 4. Interpolated map of the sediment depth (cm) of the Bisheh Zard$_1$ artificial recharge of groundwater system (after Esmaeili Vardajani, personal communication).

Table 14. Eighteen – year mean dry yield of 5 sedimentation basins and 1 recharge pond of the Bisheh Zard$_1$ ARG system, kg ha^{-1}.

Location	Shrubs	Forbs	Grasses	Perennials	Annuals	Total
Sedimentation basin No.1	44.38	48.60	407.00	44.38	455.60	500.00±150.78*
Sedimentation basin No.2	65.50	125.95	235.35	65.50	361.30	426.80± 95.28ns
Sedimentation basin No.3	152.80	76.13	187.38	152.80	263.50	416.30± 87.89ns
Sedimentation basin No.4	142.53	124.98	164.70	142.53	289.68	432.20± 96.19ns
Sedimentation basin No.5	128.58	144.40	103.33	128.58	247.73	376.30± 147.09ns
Recharge pond	107.23	163.53	107.90	107.23	271.43	378.70± 97.02

*The difference between the first sedimentation basin and the recharge pond is significant at the 5 % level.

SUMMARY AND CONCLUSIONS

The well-being of transhumant pastoralists depends substantially on maintaining productive herds of livestock. The livestock is the capital of such people on foot, and the number of healthy animals in a herd is an indication of wealth and the status symbol for the owner. Forage, water and shade are the three pillars of a sound animal husbandry in hot deserts.

The forced sedentarization of nomad herders in the 1930s, the environmentally disastrous land reform during the 1960s, and finally, the self-sufficiency policy in wheat and red meat in the late 1970s have wreaked havoc on our soil and groundwater resources. These failed strategies accelerated desertification and caused city-ward migration for millions of mixed farmers and nomads with dire consequences.

Drought is an ever present threat in drylands, and for places where most or all of their water needs is supplied through underground resources, the drastic recession of watertable is tantamount to drought (Kassas, 1987). Thus, the over-exploiters of groundwater create a permanent drought for themselves. Implementation of aquifer management and its linchpin, the ARG, successfully alters this critical situation.

Most deserts of the world are endowed with potential aquifers with truly gigantic capacities. Moreover, occurrence of intense rains on the denuded desert watersheds cause flush floods with unfortunate consequences. The floodwaters should be stored in aquifers and consumed prudently. Aquifer management, careful planning and efficient use of the available resources, particularly the groundwater, mitigate the effects of prolonged droughts. *This is desertification control par excellence!* This is a variation on the theme of *Genesis Strategy* with the difference that here water is saved under ground and consumed commensurate with needs. Collection and preservation of hay in rainy years to carry herds during the future droughts is out of the question.

The higher yield ratio of the spate-irrigated rangeland to that of control during droughts in the Gareh Bygone Plain, 2.86 vs. 2.69 fold in wet years points to the less obvious positive effect of spate irrigation of rangelands. This affirms the conclusion reached by Schlesinger et al. (1990) and supportive references that episodic availability of soil resources, especially water, leads to a "pulse reserve" response of desert ecosystem. Abundance of geologic nitrogen-enriched floodwater during wet years (Yazdian and Kowsar, 2003) facilitates soil nutrient turnover that improves and enhances plant growth in droughts. To this may be added the relatively high concentration of organic phosphorus in the litter fall of the spate-irrigated eucalyptus trees, which spread over the ARG system (Khanmirzaei et al., 2009). Thus, the possible N×P interaction facilitates a higher forage yield in such systems. Although the water demand of control plots during the growing season of wet years may be met by the timely and sufficient rainfall, the rather low available water capacity of the loamy sand causes the drainage of part of the rain water below the rooting zone, thus ineffectual for primary production, while the loam-sandy loam soil of the spate-irrigated plots retain a much higher volume of water even in the absence of flooding.

Floodwater spreading for the artificial recharge of groundwater since January 1983 on 2445 ha of the GBP has converted a substantial part of that degraded land into arable land. The satellite imageries of 1984, 1998, and 2002 reveal that in addition to the flood irrigated systems, the area of poor and very poor rangeland has decreased by 827 ha. The mean annual increase in the yield of allowable use of forage due to floodwater irrigation has been 261.93 kg per ha. Assuming that the daily intake of an average weight livestock (30.0 kg) is 1.5 kg, the 2070 ha that receive good watering can annually support 990 heads of livestock over and above the number which graze the untreated rangeland. Moreover, the stubble produced in an additional 1000 ha of irrigated fields in an average year, supply stubble for an extra 9132 heads of sheep and goats. This, and the extra income from the irrigated agriculture, has placed most of the mixed farmers above the poverty line.

A better indicator for the effectiveness of spate irrigation of a rangeland is an improvement in animal performance, with the subsequent increase in the kilogram of meat, milk, wool, hide and dung produced per ha relative to the control. However, the secretiveness of the settled nomads, which has been ingrained in them through the centuries of evading heavy-handed tax collectors and marauding bandits, deprived us of obtaining accurate data on the jump in their income due to the accomplished activities. Therefore, we have to rely on anecdotal evidence and rumors: unheard of double-kidding and lambing occurred in the 1984-85 breeding season and thereafter; purchasing and renewing cars and pickup trucks has become a yearly ritual for the formerly disadvantaged herders; a nearly 1000-fold jump in the price of farm fields has increased the land rush and the number of court cases related to expropriation of public rangelands.

As envisioned at the initial planning of the project, supplying water was the most important action that the Government could undertake to convince the environmental refugees to return home. Reverse migration since 1984 has increased population of the 4 villages that surround the ARG systems. We may safely claim that a rare phenomenon is happening here; a jump in population is mainly due to the addition of middle-aged people, not the baby boomers.

A substantial rise in income has brought prosperity to the majority of nomads. As they annually sell their surplus wheat grain to the Government, and this requires improved infrastructure, 20 km of paved road have been constructed for the villages. Furthermore, they have been provided with safe water and connected to the electricity and communication grids.

Forage availability in late summer and fall is an important concern of stockmen in drylands. Stall-feeding, usually with barley grains, is an expensive enterprise. Therefore, most stockmen have to undersell their livestock in fall. We let them graze the ARG systems at 3% cost of stall-feeding. This money is spent on removing the bed load and coarse sediments from the canals and channels, and minor repairs of the systems.

Valentine (1971) recommended fertile loams as the most desirable soil for spate irrigation. However, he contended that all soils, except the gravelly, deep, sandy, saline, or shallow soils on bedrock were suitable for this irrigation technique.

We have shown that contrary to Valentine's conclusion, it is possible to reclaim practically a denuded sandy expanse and provide livelihoods for a group of disadvantaged people using an appropriate, indigenous technology at a very low cost. An improvement in soil quantity and quality and biodiversity, an increase in carbon sequestration, and mitigation of flood damages and mortalities are the side benefits of aquifer management. Besides being technically practicable, socially acceptable, and environmentally sound, aquifer management is financially viable. In fact, the benefit: cost ratio for this project has been estimated at 22:1, if the intangible benefits and flooding fatalities are not considered (Bakhtiar et al., 1997). Karimzadegan et al. (2001) employed the contingent valuation method (CVM) and the extended linear expenditure system (ELES) to assess the nonuse values of the Gareh Bygone man-made forest. The sum of the benefit: cost ratios for the three principal non-market services of the forest, consisting of climate regulation, soil erosion control, and habitat/ wildlife refuge was 83:1.

Breaking the non-arable land for dry farming not only has encroached on the traditional grazing ground of nomads, but also has expanded the severely degraded rangeland. Seeding a proper mixture of adapted grasses and legumes on abandoned farm fields is therefore recommended, as they need the minimum of preparation (Stoddart et al., 1975).

Assuming that the global climate change would adversely affect drylands, we propose diversification of our agriculture. We emphasize raising camels for earning livelihoods in place of irrigated-agriculture. This change in occupation raises the higher value added income of the transhumant herders who are not keen farmers. Spate irrigation of indigenous drought resistant shrubs and bushes seems to be more beneficial than irrigation of cereals using groundwater. Camels easily remove forage from thorny shrubs and bushes, and they do not require daily watering Assessment of water productivity of spate-irrigated ranges for camels requires a thorough research.

There are numerous potential places for replication of this project. Out of 43 mha of alluvial fans and debris cones in the Land of Iran, with a potential to safely store 5,000 km^3 of water, 11 times the mean annual precipitation of this country, 14 mha have been proposed for development by utilizing some 42 of the 50^+ km^3 of annually wasted floodwater. *We know of no better way to make wastelands bloom!*

ACKNOWLEDGMENTS

We cannot appropriately thank the numerous colleagues from the Government of the Islamic Republic of Iran for their continuous support and encouragement. Were it not for their wholehearted cooperation we would not have a success story to report. We are also indebted to the Azad Islamic, Ghent, Gorgan, Putra, Shiraz, Tarbiat Modarres, Tehran and Wageningen Universities, the K.U. Leuven, and the ITC for encouraging one or more of their graduate students to implement their research at the Kowsar Floodwater Spreading and Aquifer Management Research, Training and Extension Station, GBP, I.R.Iran. Fereidune Arjmand, Ali Akbar Ashrafi, and Fathallah Ebrahimi of the Fars Office of Natural Resources and Watershed Management are sincerely thanked for the quail bush planting. The following colleagues from Fars Research Center for Agriculture and Natural Resources are thanked for their timely cooperation: Mr. Gholamreza Ghahhari has provided the rainfall, flooding and grazing data; Ms. Ladan Jowkar has performed the statistical analysis; Dr. Masoud Nejabat has provided the satellite-assisted imagery data on land use in the GBP. The following colleagues have participated in plant sampling and species identification: Maryam Amiri, Maryam Fakhraei, Hossein Hashemi, Ahmad Hatami, Effat Jaafari, Asghar Mehrbakhsh, Gholam Nekouian, Babak Niazi Ardekani, Mansour Niazi Ardekani, Reza Samanian, Behrooz Shadkam,and Mohammad Tayyebi.

REFERENCES

Adeel, Z.; Dent, D.; Dobie, Ph.; Mesrmann, Ch.; Niamir-Fuller, M.; Quatrini, S.; Sokona, Y. 2009. *Revitalizing the UNCCD*. The United Nations University. p.1-5.

Adeel, Z.; King, C.; Schaaf, Th.; Thomas, R.; Schuster, B. 2008. *People in marginal drylands: Managing natural resources to improve human well-being*. The United Nations University.p.12.

Bakhtiar, A.; Najafi, B.; Kowsar, A.; Habibian, S.H. 1997. Profitability and flexibility of floodwater spreading systems in Iran. *In:* B. Aminipouri and J. Ghoddousi (ed.) *Proceedings of the 8th International Conference on Rainwater Catchment Systems.* 25-29 Apr. 1997, Tehran, I.R. Iran. p.161 - 168.

Breman, H.; deWit, C.T. 1983. Rangeland productivity and exploitation in the Sahel. *Science (Washington, DC) 221:*1341-1347.

Corens, P. 2005. *The effect of floodwater spreading on the field soil water balance in the Gareh Bygone Plain*, I.R. Iran. M.S. thesis, Katholic Universiteit Leuven.p.46.

Coughenour, M.B.; Ellis, J.E.; Swift, D.M. ; Coppock, D.L.; Galvin, K.; McCabe, J.T. ; Hart, T.C. 1985. Energy extraction and use in a nomadic pastoral ecosystem. *Science (Washington, DC) 230*:619-625.

Davis, A.M. 1972. Selenium accumulation in a collection of *Atriplex* species. *Agronomy Journal 64*:823-824.

Davies, J. 2008. Viewpoints:" Are poverty and land degradation inevitable in desert-prone areas?" *Natural Resources Forum 32*:79.

EsmaeiliVardanjani, N.; Gabriels, D.; Kowsar, S.A.; Raes, D. No date. Assessment of sampling location effect on sediment mapping and estimation of sediment volume in a floodwater spreading system of the Gareh Bygone Plain, I.R.Iran (Manuscript in preparation).

Fallah Shojaei, J. 2005. *The effects of four acacia species on some physical and chemical properties of soils of the Gareh Bygone Plain,* Fasa. Unpublished M.S. thesis, College of Agriculture, Shiraz University (in Farsi).

FAO, 2005. *Global Forest Resource Assessment 2005 – Progress Towards Sustainable Forest Management*, FAO Forestry Paper 147, Food and

Agriculture Organization of the United Nations (FAO), Rome, Italy.

Farhoudi, G. 1987. A comparison of different landforms resulting from neotectonics in western and southern Iran. *Zeitschrift für Geomorphologie 63*: 57-60.

James, G.A.; Wynd, J.G. 1965. Stratigraphic nomenclature of Iranian Oil Consortium agreement area. *American Association of Petroleum Geologists Bulletin 49*: 2182-2245.

Karimzadegan, H.; Rahmatian, M.; Mahmoodi, M. 2001. Valuation of environmental assets of the Gareh Bygone man-made forest by using extended linear expenditure system (ELES) model. *Journal of Environmental Studies 26*:51-59 (in Farsi).

Khanmirzaei, A.; Adhami, E.; Kowsar, S.A.; Sameni, A.M. 2009.Organic and inorganic forms of phosphorus in a calcareous soil planted with four species of eucalyptus in southern Iran. *Communications in Soil Science and Plant Analysis 40:*3194-3210.

Kassas, M. 1987. Drought and desertification. *Land Use Policy 4*:389-400.

Kowsar, A. 1991. Floodwater spreading for desertification control: An integrated approach. *Desertification Control Bulletin 19*: 3-18.

Kowsar, S.A.2005. *Abkhandari* (aquifer management): A green path to the sustainable development of marginal drylands. *Journal of Mountain Science 2*:233-243.

Kowsar, S.A. 2007. Report of SUMAMAD activities at the project site, Gareh Bygone Plain, I.R. Iran. *In* Sustainable Management of Marginal Drylands SUMAMAD. Fifth Project Workshop, Aleppo (Syria), 12-17 November 2006. p. 64-68.

Kowsar, S.A. 2008a. Desertification control through floodwater harvesting: The current state of know-how. *In* Cathy Lee and Thomas Schaaf (ed.): The Future of Drylands. International Scientific Conference on Desertification and Drylands Research. Tunis, Tunisia, 19-21 June 2006. Man and Biosphere series. UNESCO Publishing, Paris. p.229-241.

Kowsar, S. A. 2008b. Viewpoints: "How can sustainable land management contribute to mitigating climate change?" *Natural Resources Forum 32*:254.

Kowsar, S.A. 2008c.Gareh Bygone Plain. *In* Cathy Lee and Thomas Schaaf (ed.): *Using Science to Promote Sustainable Development. Project Findings from Northern Africa to Asia. Sustainable Management of Marginal Drylands.* UNESCO, Paris. p.104-125.

Kowsar, S.A.; Pakparvar, M. 2004.Assessment methodology for establishing an Aquitopia, Islamic Republic of Iran. *In* Thomas Schaaf and Cathy Lee (ed.) Sustainable Management of Marginal Drylands SUMAMAD. *Proceedings of the Second International Workshop, Shiraz, Islamic Republic of Iran,* 29 N0vember-2December 2003. p.40-55.

Miller, R.F.; McQueen, I.S.; Branson, F.A.; Shown, I.M.; Buller, W.M. 1969. An evaluation of range floodwater spreaders. *J. Range Manage. 22*:246-257.

Naderi, A.A.; Kowsar, S.A.; Sarafraz, A.A. 2000. Reclamation of a sandy desert through floodwater spreading: I. Sediment–induced changes in selected soil chemical and physical properties. *Journal of Agricultural Science and Technology (Iran) 1*:9-20.

National Oceanic and Atmospheric Administration and University of Missouri-Columbia. 1979. A study of the Caribbean Basin drought/food production problem: Final report. U.S. Department of State, University of Missouri, Columbia.

Nautiyal, S.; Rao, K.S. ; Maikhuri, R.K.; Saxena, K.G. 2003.Transhumant pastoralism in the Nanda Devi Biosphere Reserve, India: A case study in the buffer zone. *Mountain Research and Development 23*:255-262.

Pooladian, A.; Kowsar, S.A. 2000. Aquifer management: A prelude to reclamation of salinized soils. *Desertification Control Bulletin 36*: 78-82.

Rahbar, G.R.; Kowsar, S.A. 1997. Infiltrability enhancement in sedimentation basins by sowbugs. *In*: : B. Aminipouri and J. Ghoddousi (ed.) *Proceedings of the 8th International Conference on Rainwater Catchment Systems.* 25-29 Apr. 1997, Tehran, I.R. Iran.p.169 - 172.

Rahbar, G.R.; Kowsar, S.A; Jowkar, L. 2007.Bioassay of the sowbugs' burrowed material using barely and wild rye. *In* Sustainable Management of Marginal Drylands SUMAMAD. Fifth Project Workshop, Aleppo (Syria) 12-17 November 2006. p. 127-130.

Rahbar, G.R.; Kowsar, S.A.; Khalili, D. 2001. Soil modification by the freshly-laid sediment and colonies of sowbug (*Hemilepistus shirazi* Shuttz) in a sandy desert. *Proceedings of the First Congress of Soil Scientists of Tajikistan*, 2-3 November 2001, Dushanbeh, Tajikistan. p. 37-39.

Soil Survey Staff. 1999. Keys to soil taxonomy. USDA, NRCS, 8th edition. US GPO, Washington (DC). p.353.

Stoddart, L.A.; Smith, A.D. 1955. *Range management* (2nd edition). New York, NY: McGraw-Hill Book Company.p.366, 384.

Stoddart, L.A.; Smith, A.D.; Box, Th. W. 1975. *Range management* (3rd edition. New York, NY: McGraw-Hill Book Company.p.9, 169, 483.

Tannehill, I.R. 1947. *Drought, its causes and effects.* Princeton, NJ: Princeton University Press, p. 113.

UNDP. 2003. The global dryland imperative. Distributed at the 7th International Conference on Development of Dry Lands, 14-17 Sept. 2003, Tehran, I.R.Iran.

UNEP. 2007. Global environmental outlook. GEO-4. United Nations Environment Programme (UNEP). Nairobi, Kenya.

Valentine, J.F. 1971. *Range development and improvements*. Brigham Young University Press, Provo, UT. p. 318

Vetaas, O.R.; Knudsen, A. 2004. Fragile mountains-Fragile people? Understanding "fragility" in the Himalayas. Report on a workshop in Norway. *Mountain Research and Development 24*:182-183.

Wolman, M.G. 1985. Soil erosion and crop productivity: A worldwide perspective. *In* R.F. Follet and B.A. Stewart (ed.) Soil Erosion and Crop Productivity. ASA, CSSA, and SSSA Publishrs, Madison, WI.

WRI, 1998. *World Resources 1998-99: Environmental Change and Human Health*,World Resources Institute (WRI), Washington DC, USA.

Yazdian, A.R.; Kowsar, S. A. 2003. The Agha Jari Formation: A potential source of ammonium and nitrate nitrogen fertilizers. *Journal of Agricultural Science and Technology (Iran) 5*:153-163.

Young, J.A.; Kay, B.L.; George, H.; Evans, R.A. 1980. Germination of three species of *Atriplex*. *Agronomy Journal 72*: 705-709.

ACRONYMS

AJF	Agha Jari Formation
ARG	artificial recharge of groundwater
AUC	allowable use coefficient
AWC	available water capacity
CVM	contingent valuation method
ELES	extended linear expenditure system
FAO	Food and Agriculture Organization of the United Nations
GBP	Gareh Bygone Plain
ILWIS	integrated land and water information system
ITC	International Institute for Aerospace Survey and Earth Sciences
km	kilometer
MAP	mean annual precipitation
ha	hectare
mha	million hectares
mm	millimeters
mm^3	million cubic meters
UNDP	United Nations Development Programme
UNEP	United Nations Environment Programme
UNESCO	United Nations Educational, Scientific and Cultural Organization
WRI	World Resources Institute

In: Agricultural Production
Editors: Felix C. Wager, pp. 79-92

ISBN: 978-1-61668-695-6

Chapter 3

A STOCHASTIC EVALUATION OF THE VALIDITY OF AN ANIMAL-HEALTH SURVEY IN URUGUAY

***K. Suzuki**[*1], **C. Galosi**[1, 2], **G. Echeverría**[1, 3], **D. Satragno**[1, 4], **F. Alzugaray**[1, 4], **E. de Torres**[1, 4], **H. Guarino**[1, 4] **and A. Freyre**[1, 4]*

[1]Universidad Nacional de La Plata, La Plata, Argentina
[2] Comisión de Investigaciones Científicas de la Provincia de Buenos Aires (CIC), La Plata, Argentina
[3] Instituto de Genética Veterinaria "Fernando Noel Dulout" - Facultad de Ciencias Veterinarias - Centro Científico Técnico de La Plata - Consejo Nacional de Investigaciones Científicas y Técnicas (IGEVET-FCV-CCT La Plata CONICET), La Plata, Argentina
[4]Universidad de la República, Montevideo, Uruguay

ABSTRACT

This paper introduces spreadsheet simulation models to evaluate the validity of national or regional disease surveys aimed at identifying infection in populations of farm animals. The process of evaluation includes specification or calculation of cluster-level diagnostic test sensitivity (the proportion of animals with the disease which test positive) and specificity (the proportion of animals without the disease which test negative), which are obtained from two probability distributions of the number of positive tests at individual-level expected from infected and non-infected clusters, respectively.

* Corresponding author: Tel./ fax: +54 221 425 3276; E–mail address: pvs@provetsur.net (K. Suzuki). PROVETSUR, Facultad de Ciencias Veterinarias, Universidad Nacional de La Plata, Avda. 60 y 118, La Plata, C.P. 1900, Argentina

Probability distributions for the number of positive clusters expected in a situation of the herd prevalence are specified and used to define survey properties (the survey being considered a diagnostic system), and receiver operating characteristic curves (consisting of a plot of sensitivity and specificity pairs for different cut-off values) are drawn. The result of a survey implemented to determine the prevalence of infectious bovine rhinotracheitis in dairy cattle in Uruguay, South America was used to illustrate this approach. The models can be adapted to a wide range of survey designs in animal health and production.

INTRODUCTION

The necessity to supply valid data on the disease status of an animal population in a given country or region is becoming more important, with the opening of the world market in the context of the World Trade Organisation. As a result, the epidemiological methods used and the interpretation of laboratory diagnostic test results will be open to examination. When making sure that an animal population in a country or a region is free from infection or disease, it is basically more crucial to identify the supposed clusters of infected animals than infected animals themselves. When a number of animals are being examined within clusters to specify the infection status of each cluster, the testing regimen should be assessed as a cluster-level test. The regimen is characterised by its sensitivity (the proportion of animals with the disease which test positive) and specificity (the proportion of animals without the disease which test negative), being dependent on some criteria incorporating the prevalence of infected animals within infected clusters, the cluster size, the number of animals tested, the characteristics of the animal-level test used and the cut-off number of positive animal-level tests chosen to proclaim a cluster as infected [1]. In addition, It was reported that sensitivity and specificity at cluster-level would be influenced by differences in test sensitivity and specificity at animal-level between clusters [2]. Subsequently, A simulation model was made to assess how cluster-level sensitivity differs with test characteristics at animal-level, within-cluster prevalence of infection, and sample size [3]. Over recent years, sample-size calculations in the context of surveys intending to prove freedom from infection or disease have depended on formulas and tables [4]. Significant presumptions for these calculations were that the sampled population was infinite, and that the testing procedure utilised was perfect (i.e. with a sensitivity and specificity of both 100%). However, the interpretation of animal-health survey results (both to prove freedom from infection and to define the true prevalence of infected or diseased herds in a population) is not simple and must take into account their inherent variability or uncertainty. This paper discusses stochastic simulation models that can be used to evaluate the validity of such surveys and to provide help for the interpretation of screening test results. The results of a survey implemented to estimate the prevalence of infectious bovine rhinotracheitis (IBR) in Uruguay are used to show the approach.

Materials and Methods

The models has been made for the evaluation of the validity of surveys where a preset number of animals (the individual units within clusters of animals) is examined using a screening test (which may be a combination of animal-level tests) to determine if the clusters are infected or diseased. A number of clusters is examined in order to state that a country or region is free from infection or disease, or to identify that the infection or the disease arises. Therefore, the survey itself is regarded as a screening process or a diagnostic system. In this paper, a cluster can be any group of individual animals such as a herd, a flock, a mob or a group in a pen (the word "herd" is basically used to describe the models hereinafter). In addition, the authors will make reference only to "infections" and "countries", but the reader will comprehend that the concepts also apply to diseases and regions, respectively. Infections as defined in this paper are caused by the presence of a contagious agent within an animal; the infection might cause only subclinical transformations such the development of antibodies or might cause the development of clinical signs (i.e. disease).

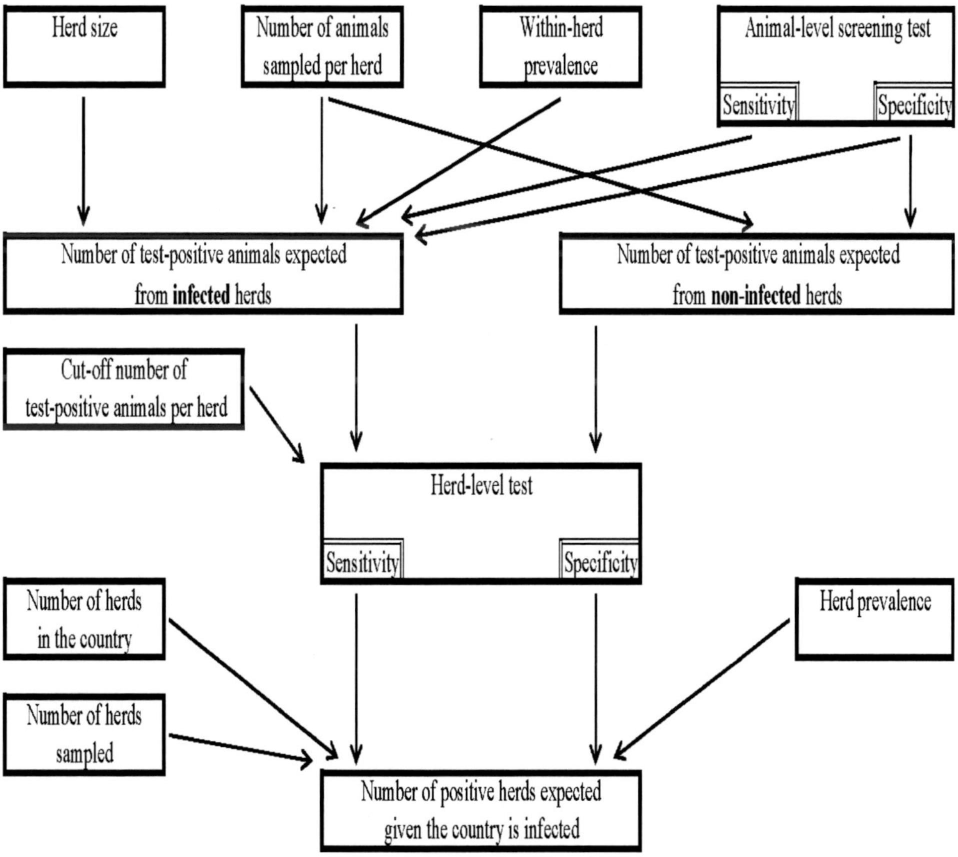

Figure 1. Structure of the simulation model.

Structure of the Simulation Model and Overall Process of Evaluation

The process of evaluation includes two steps; the specification of herd-level test sensitivity and specificity, and the specification of these properties at the country level. Therefore, this paper will bring the reader to three consecutive level of consciousness: the animal level, the herd level, and the country level (Figure 1). The formulas and probability distributions used in the models are shown in Table 1. The models were written in Microsoft Excel 2007 (Microsoft Corporation, Redmond, WA, USA) and simulated using @Risk version 5.0.1 (Palisade Corporation, Newfield, NY, USA), a software allowing Monte Carlo simulation with Latin Hypercube sampling.

Number of Positive Animals Expected from Infected and Non-Infected Herds and Herd-Level Test Characteristics

The screening test results applied to a number of animals within a herd can be defined using the concepts of test sensitivity and specificity at herd-level [1]. Herd-level test sensitivity was interpreted as the probability of observing a number of test-positive results at animal-level (which may contain false-positive results) equal to or above a predetermined cut-off number, given that the herd is really infected. Herd-level test specificity was interpreted as the probability of observing a number of test-positive results at animal-level (which are all considered to be false-positive results) below this cut-off number, given that the herd is really non-infected. To evaluate a test at the animal level, its characteristics are determined from the distribution of test results [e.g. an optical density or percentage of reactivity for an enzyme-linked immunosorbent assay (ELISA)] from truly infected and truly non-infected animals. Likewise, herd-level test characteristics are determined from the distribution of results (i.e. the number of animals positive to the animal-level test) from truly infected and non-infected herds. As illustrated in Figure 1 and described in Table 1, these distributions depend on the following parameters:

- For truly infected herds: the number of infected and non-infected animals sampled per herd and the sensitivity and specificity of the animal-level screening test. The number of infected animals sampled from an infected herd depends on the size of the herd, the prevalence of infection within that herd, and the total number of animals sampled per herd.
- For truly non-infected herds: the number of animals sampled per herd and the specificity of the animal-level screening test.

Stochastic modelling allows the derivation of these two probability distributions from the parameters defined above and their inherent variability or uncertainty (Table 1).

- Most laboratories report sensitivity and specificity for diagnostic tests. When validation data to define these characteristics are available, Beta distributions can be used to model probability distributions for true sensitivity and specificity. The Beta distribution models the probability of a unit (e.g. an animal) randomly recruited from a population having a particular characteristic such as being infected (i.e. a binomial event) if a number of units within a randomly recruited group of units has been found with the characteristic [5].

- National animal-population statistics are used to model herd size, and the most appropriate parametric or non-parametric probability distribution fitting these data is utilised. If these data are given as the number of herds per interval of herd sizes (i.e. as an histogram), they can be modelled using non-parametric distributions such as the histogram distribution function, or the discrete distribution of @Risk as in the IBR example presented in this paper. A combination of uniform and discrete distributions also can be used.
- The within-herd prevalence of infection is modelled using a RiskPert distribution with data chosen based on expert opinion and a review of published literature. The RiskPert distribution is believed one of the most-suitable distributions to model a continuous variable for which experts have been asked to provide three values (i.e. a minimum, a maximum, and a most-likely value) [5].
- The number of animals sampled per herd (N) is a fixed value. In this study the authors used the two different fixed values for N, such as five (model 1) and 10 (model 2). The authors compared these two models to select the better value for N. It is our thought that this model may be utilised in the planning stage of a survey to calculate that sample size, but it will be beyond the scope of this paper.
- The number of infected animals expected in the herd sampled is modelled using a hypergeometric distribution with parameters accounting for the number of animals sampled per herd, the number of infected animals in the herd, and the size of each herd. The number of infected animals in the herd is simply obtained from the value of the product of the herd size with the within-herd prevalence rounded to the unit with a minimum of one infected animal. The hyper geometric distribution is considered the most suitable as sampling is done without replacement. In situations where sampling fractions are low (e.g. below 10%), the hypergeometric distribution could be approximated by the binomial distribution with parameters explaining the number of animals sampled from each herd and the within-herd prevalence [6]. Thus, they are restricted by the minimum and the maximum value, which may be different from 0 and 1.

The two probability distributions are produced from a first-stage simulation with 1000 iterations. The 1000 values obtained for each distribution are saved into an Excel spreadsheet where the superimposed probability distributions (i.e. the relative frequency distributions of output values) can be drawn. These results provide point estimates of test sensitivity and specificity at cluster-level for each possible cut-off number of positive animal-level tests, and a receiver operating characteristic (ROC) curve is also drawn, and an area under the ROC curve (AUC) is calculated. The ROC curve shows the characteristics of a diagnostic test by graphing the false-positive proportion (1 - specificity) on the horizontal axis and the true-positive proportion (sensitivity) on the vertical axis for various cut-off numbers. The AUC is a popular measure of the accuracy of a diagnostic test. Other things being equal, the larger the AUC, the better the test is a predicted existence of the disease. The possible values of AUC range from 0.5 (no diagnostic ability) to 1.0 (perfect diagnostic ability). These calculations are based on the presumption that the distributions of test-positive animals in both the infected and non-infected populations of herds are similar to the sample distributions (the distributions of 1000 values).

The cut-off number of test-positive animals is selected from the above probability distributions and ROC curve depending on the objective of the survey. For example, sensitivity at herd-level could be maximised to give high confidence in negative results at herd-level when attempting to substantiate freedom from infection.

Table 1. Description of model inputs for the example of IBR in Uruguay

Input variables	Notation	Formula used with @Risk (Microsoft Excel notation used)
Animal-level screening test characteristics		
Herd size	Size	RiskDiscrete (distribution of herd size for 19 departments, proportion of herds associated with each department)
Number of animals sampled per herd	N1	Fixed values ["5" for model 1 and "10" for model 2]
Within-herd prevalence	Prev	RiskPert (0.3, 0.7, 1.0)
Test sensitivity at animal-level	Ase	RiskPert (0.94, 0.98, 0.995)
Test specificity at animal-level	Asp	RiskPert (0.815, 0.97, 0.985)
Herd-level test characteristics		
Number of infected animals in a corresponding herd	Infl	IF (ROUND (Prev * Size, 0) = 0, 1, ROUND (Prev * Size, 0))
Number of infected animals expected in the herd sampled	Inf2	RiskHypergeo (N1, Infl, Size)
Number of test-positive animals expected from infected herds	Pos	SUM (IF (Inf2 > 0, RiskBinomial (Inf2, Ase), 0): IF (N1 - Inf2 > 0, RiskBinomial (N1 - Inf2, 1- Asp), 0))
Number of test-positive animals expected from non-infected herds	Neg	RiskBinomial (N1, 1- Asp)
Number of simulations used to derive the probability distributions of "Pos" and "Neg"	Iter	Fixed value [1000]
Number of simulations giving a value of "Pos" more than or equal to a cut-off number C	Cpos	
Number of simulations giving a value of "Neg" less than a cut-off number C	Cneg	
Test sensitivity at herd-level	Hse	RiskBeta (Cpos + 1, Iter - Cpos + 1)
Test specificity at herd-level	Hsp	RiskBeta (Cneg + 1, Iter - Cneg + 1)
Country-level test characteristics		
Number of herds in the country	Pop	Fixed value [51,072]
Number of herds sampled	N2	Fixed value [14]
Herd prevalence	Hprev	RiskPert (0.40, 0.57, 0.74)
Number of infected herds in a corresponding population	Inf3	IF (ROUND (Hprev * Pop, 0) = 0, 1, ROUND (Hprev * Pop, 0))
Number of infected herds expected in the population sampled	Inf4	RiskHypergeo (N2, Inf3, Pop)
Number of positive herds expected given the country is infected	Pos2	SUM (IF (Inf4 > 0, RiskBinomial (Inf4, Hse), 0): IF (N2 – Inf4 > 0, RiskBinomial (N2 – Inf4, 1- Hsp), 0))

Number of Positive Herds Expected and Survey Properties

The next section of the spreadsheet models is used to approximate the properties of the survey (i.e. its sensitivity and specificity) and AUCs. Survey (or country-level) sensitivity is determined as the probability of proclaiming the country as infected, given that the infection exists. Survey (or country-level) specificity is determined as the probability of proclaiming the country as free from infection, given that it really is. Survey properties are determined from results (i.e. the number of positive herds to the herd-level test) of two groups of a large number of surveys conducted in the same country using the same population of herds. For one group of surveys (Group 1), one thinks that the country is truly infected while for the other (Group 2), one thinks that the country is truly free from infection. As illustrated in Figure 1 and described in Table 1, these distributions are obtained from the following criteria:

- Group 1: The number of infected and non-infected herds sampled in the survey and the test sensitivity and specificity at herd-level.
- Group 2: The number of herds sampled and the test specificity at herd-level.

These two probability distributions of the number of positive herds expected are obtained from the parameters defined above and their inherent variability or uncertainty (Table 1).

- Test sensitivity and specificity at herd-level are modelled as Beta distributions using results from first-stage simulation data, similar to the way of animal-level screening test characteristics modelled from laboratory validation data.
- The herd prevalence depends on the epidemiology of the infection concerned within the considered country, that is:

For infections that do not transmit (or that are not likely to transmit) quickly between herds (such as bovine leucosis or bovine tuberculosis), the herd prevalence is fixed at a threshold minimum level under which one would accept to regard the country as "free from infection". For infections that are known to transmit quickly between herds after introduction into a country (such as in the IBR example demonstrated in this paper) the herd prevalence is modelled to give a more-realistic scenario. As for the within-herd prevalence defined formerly, the herd prevalence is modelled using a RiskPert distribution with maximum, minimum and most-likely values selected based on expert opinion or a review of published literature [5].

- The number of herds sampled for the survey (N2) is a fixed value. The calculation of this sample size will not be further developed in this paper.
- The number of infected herds expected in the population sampled, if the country is infected, is modelled using a hypergeometric distribution with parameters accounting for the number N2, the number of infected herds in a corresponding population, and the number of herds in the country. The number of herds in a corresponding population is simply obtained from the value of the product of the number of herds in the country with the herd prevalence rounded to the unit with a minimum of one infected herd. The number of herds in the country (i.e. population size), however,

could be omitted if the sampling fraction of clusters is low (e.g. below 10%). In this situation, the hypergeometric distribution could be approximated by the binomial distribution (i.e. sample with replacement) [6].

Probability distributions of the number of positive herds expected are obtained from a second-stage simulation with 1000 iterations. The model lets one to acquire such distributions for a range of input herd prevalence entered deterministically (fixed minimum threshold value) or stochastically (modelled realistic values). The spreadsheet models are used to draw superimposed probability distributions, obtain survey sensitivity and specificity for each input herd prevalence and possible cut-off number of positive herds, and respective ROC curves.

The Example Survey: IBR Survey in Uruguay

A serological animal-health survey conducted in Uruguay between November 2008 and May 2009 was used to evaluate the validity of the spreadsheet simulation models. For this survey, the presumptions and several model parameters were as follows:

- Blood samples from 90 different cattle were collected from each of the 14 study herds. However, for the sake of illustration, the numbers of animals sampled per herd in the spreadsheet simulation models were determined as five and 10 for the models 1 and 2 mentioned above, respectively.
- Sera were analysed using a commercial indirect ELISA (Chekit® Trachitest IBR Antibody Test Kit, IDEXX, Switzerland). The published test sensitivity and specificity were modelled using a RiskPert distribution with minimum, most-likely, and maximum values being 94%, 98% and 99.5% for sensitivity, and 81.5%, 97% and 98.5% for specificity, respectively [7, 8].
- Herd sizes were modelled using 2008 statistics for cattle herds in Uruguay obtained from the website of Directorate General of Livestock Services of the Government of Uruguay (http://www. mgap.gub.uy/dgsg/DICOSE/DatosDJ_2008.htm). For this simulation study, herd sizes were obtained within each herd-size interval from the integer value. A discrete probability distribution was fitted to give the integer values with weights accounting for the relative frequencies of herds within each herd size interval.
- The within-herd prevalence was modelled using a RiskPert distribution with the results of the serological survey mentioned above. Minimum, most-likely, and maximum values for within-herd prevalence were 30%, 70% and 100%, respectively. While the herd prevalence was modelled using a RiskPert distribution with a review of published literature regarding a meta-analysis for IBR in Uruguay [9]. Minimum, most-likely, and maximum values for herd prevalence were 40%, 57% and 74%, respectively.

RESULTS

Descriptive statistics for a selection of key output variables for the IBR survey are shown in Table 2.

Table 2. Descriptive statistics of animal- and herd-level test characteristics, and numbers of positive herds expected by the models from an IBR survey in Uruguay

Inputs/outputs	Minimum	5th percentile	Mean	95th percentile	Maximum	SD
Animal- and herd-level test characteristics						
Ase	0.944	0.958	0.976	0.990	0.995	0.010
Asp	0.839	0.894	0.947	0.980	0.985	0.027
[Model 1]						
Hse	0.907	0.921	0.934	0.946	0.956	0.008
Hsp	0.950	0.961	0.970	0.978	0.988	0.005
[Model 2]						
Hse	0.972	0.981	0.987	0.992	0.995	0.004
Hsp	0.965	0.975	0.982	0.988	0.993	0.004
Number of positive herds expected by the models						
Model 1	2	4	7.6	11	13	2.0
Model 2	2	5	8.0	11	14	2.0

Ase and Asp: Animal-level test sensitivity and specificity, respectively. Hse and Hsp: Herd-level test sensitivity and specificity, respectively. Model 1 and 2: the model with five and 10 animals sampled per herd, respectively (number of herds sampled = 14).

Number of Positive Animals Expected From Infected and Non-Infected Herds and Herd-Level Test Characteristics

At the animal level, the sensitivity of the ELISA used in the IBR survey ranged from 94.4% to 99.5% with a mean of 97.6%. Specificity of this test ranged from 83.9% to 98.5% with a mean of 94.7%. Output probability distributions of the expected numbers of positive animals from non-infected and infected herds for the models 1 and 2 are presented in Figures 2 and 3, respectively, while the ROC curves of the herd-level test used for the models 1 and 2 are presented in Figures 4 and 5, respectively. From these simulated data, the authors selected the cut-off numbers of positive animals under which a herd would be considered as non-infected to be 2 and 3 for the models 1 and 2, respectively. The AUCs for the models 1 and 2 were 98.4% (95% CI: 98.2-98.7%) and 99.7% (95% CI: 99.6-99.9%), respectively. In this simulation using 1000 iterations 66 iterations showed no positive animals from infected herds, and 30 iterations showed at least one positive animal from non-infected herds for the model 1. Therefore, point estimates of herd-level test sensitivity and specificity for the model 1 were 93.4% and 97.0%, respectively. While for the model 2, 13 iterations showed no positive animals from infected herds, and 18 iterations showed at least one positive animal from non-infected herds. Therefore, point estimates of herd-level test sensitivity and specificity for the model 2 were 98.7% and 98.2%, respectively. These data were used for the second-stage simulation to model the number of positive herds expected using Binomial distributions as described in Table 1.

Number of Positive Herds Expected and Survey Properties

Output probability distributions for the number of positive herds expected from an IBR survey conducted in Uruguay under the assumed prevalence of 57% (modelled between 40% and 74%) are superimposed in Figure 6. Descriptive statistics for the number of positive herds expected under the two different models mentioned above are presented in Table 2. In the context of the IBR survey, the number of positive herds expected ranged from 2 to 13 with a mean of 7.6 for the model 1. For the model 2, the number of positive herds expected between 2 and 14 with a mean of 8.0. The ROC curves of the survey (i.e. the diagnostic system) for the two different models were produced (figures not shown). As the country-level test characteristics, the model 1 had the AUC of 99.98% with the sensitivity of 99.8% and specificity of 99.1%. Meanwhile the model 2 had the AUC of 99.99% with the sensitivity of 99.5% and specificity of 99.9%. The survey appeared to be almost perfect in identifying IBR infection if country-level prevalence was 57% (modelled between 40% and 74%). Seven out of 14 study herds in this survey had prevalence of more than 57%, almost corresponding to the results of the simulation.

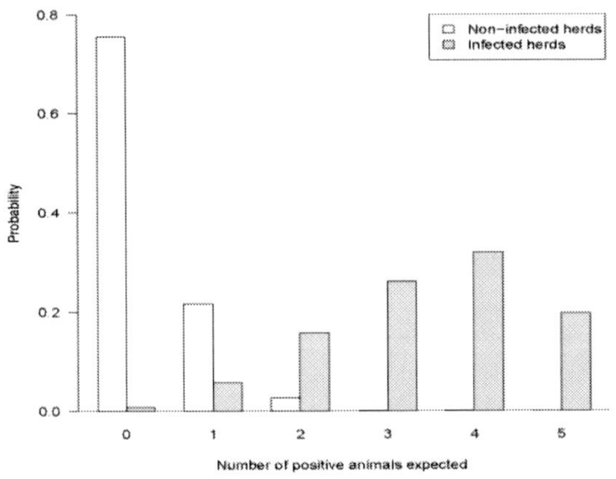

Figure 2. Output probability distributions of the number of positive animals expected against IBR amongst five animals randomly selected from non-infected herds and infected herds.

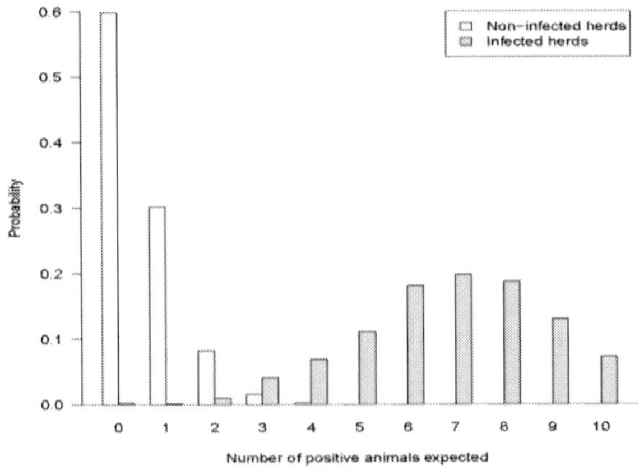

Figure 3. Output probability distributions of the number of positive animals expected against IBR amongst 10 animals randomly selected from non-infected herds and infected herds.

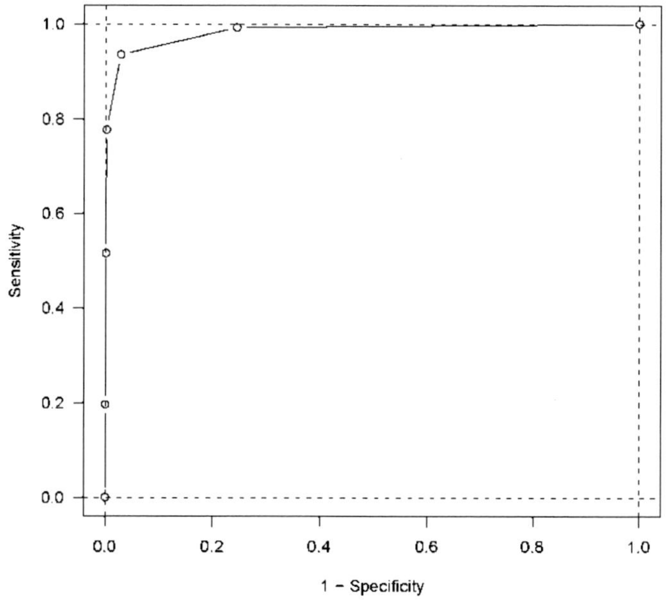

Figure 4. Output receiver operating characteristics (ROC) curve of herd-level test for seropositivity against IBR from five animals sampled per herd.

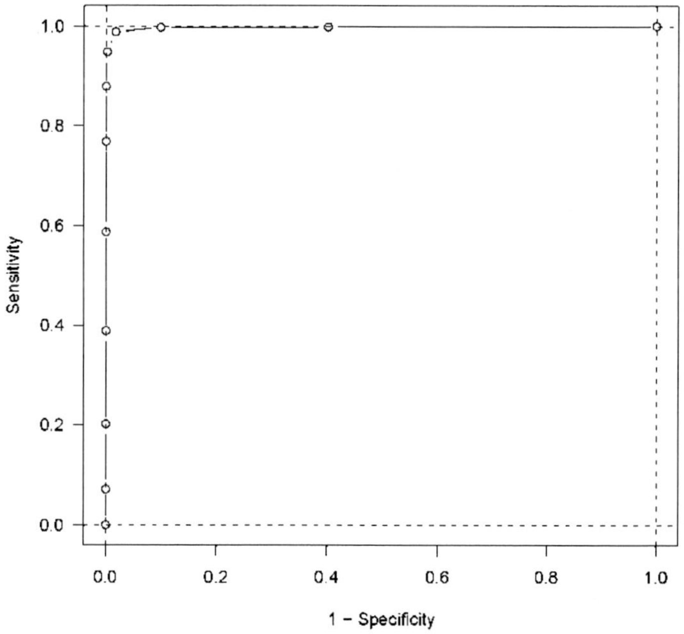

Figure 5. Output receiver operating characteristics (ROC) curve of herd-level test for seropositivity against IBR from 10 animals sampled per herd.

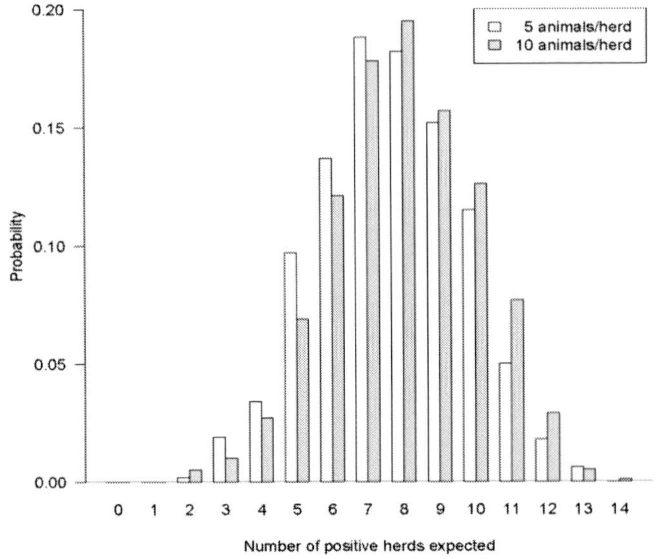

Figure 6. Output probability distributions of the number of positive herds expected against IBR amongst 14 herds sampled by the models with five and 10 animals sampled per herd, respectively.

DISCUSSION

Simulation techniques have already been used to model the cluster-level sensitivity of a diagnostic procedure [3], but it was in the context of a herd-health programme for pigs, and the process was not extended beyond the herd-level. A survey aimed at demonstrating Norway's prevalence of IBR has provided estimates of herd-level diagnostic regimen using a combination of two tests [10]. To our knowledge, this concept has not yet been discussed and reported at the country level. At the animal-level, it is important to obtain precise estimates of the characteristics of the screening test used. In our thought, Beta distributions are functional for modelling the real animal-level test characteristics using validation data from the laboratory instead of deterministic point estimates. This modelling approach has been adapted from the modelling of the true animal-level prevalence in a population [5]. If there are doubts on the validity of some characteristics, probability distributions other than the Beta can be utilised to model the uncertainty associated with these characteristics. In this example, the authors considered it appropriate to use a RiskPert distribution to model the most-appropriate values. The probability distribution used in this paper to model herd sizes has been selected to fit the national statistics but it may be changed for other surveys depending on the nature of the data. This approach considers that herds are randomly selected from the population described by these statistics - which is not truly correct when sampling is performed at farms. It is crucial to make sure that herd sizes smaller than the number of animal selected per herd be excluded from the distribution; otherwise, some iterations will be unrealistic (i.e. when more animals are selected per sample than are present in the herd). In the context of the IBR survey, herds with less than 10 cattle were not taken into account. The prevalence of infection within infected herds is modelled using the RiskPert distribution, which is considered one of the most-realistic distributions to model expert opinion [5]. When data are available from an infected country, other distributions may be used [3]. Stochastic simulations were run with the presumption that there is no relationship between the herd size and the within-herd prevalence. However, a quantification of such dependency is possible depending on the availability of data that may support it. The number of animals sampled per herd and the cut-off number of positive animals are fixed values and can be selected to optimise herd-level test characteristics. Various criteria may be used to determine the cut-off number of a test including cost considerations [11]. In the context of the IBR survey, the number of animals sampled per herd was determined prior to the development of the model using the approach proposed by Cannon and Roe (1982). Our evaluation (which in addition takes the animal-level test characteristics into consideration) shows that our choice was acceptable (Figure 2 and 3). The cut-off number of positive animals was selected to be two (for the model 1) and three (for the model 2), and resulting point estimates of herd-level test sensitivity and specificity were 93.4% and 97.0% (for the model 1), and 98.7% and 98.2% (for the model 2), respectively. To increase these sensitivity and specificity would have required increase in sample size but the survey cost considerations would be more necessary. As was true also for animal-level tests characteristics, Beta distributions are practical to model true herd-level test characteristics (instead of using the deterministic point estimates above). However, note that Beta distributions will never consider a herd-level test to be perfect even if point estimates of either or both sensitivity and specificity are 100%. Because our process uses simulated data, the correctness of the estimates depends on the number of iterations predetermined to run the model. In this paper, 1000 iterations were selected arbitrarily. The last part of the model brings the study to the country level. The process of evaluation is similar to that at the herd level. Probability distributions presented in Figure 6 are worthwhile for the evaluation of surveys, through the determination of their sensitivity and specificity.

CONCLUSION

The models have been created on simple spreadsheet models and can be adapted and utilised in the context of a wide range of other surveys used to identify infection status. The use of a standard spreadsheet allowing Monte Carlo simulation modelling also makes the process easier to a majority of people not familiar with computer programming. The current models are therefore very useful to survey managers and decision-makers.

ACKNOWLEDGMENTS

This study was carried out as part of the project for the capacity development for improvement of livestock hygiene in the southern part of South America through regional cooperation [in Spanish: Proyecto de desarrollo profesional continuo para los veterinarios del Sur (PROVETSUR)], funded by the Japan International Cooperation Agency.

Reviewed By

Dr Son Thi Thanh Dang, Department of Veterinary Disease Biology, Faculty of Life Sciences, University of Copenhagen, Denmark.

References

[1] Martin SW, Meek AH, Willeberg P. *Veterinary epidemiology*. Ames: Iowa State University Press; 1987.

[2] Branscum AJ, Gardner IA, Wagner BA, McIntuff PS, Salman MD. Effect of diagnostic testing error on intracluster correlation coefficient estimation. *Preventive Veterinary Medicine*. 2005 Jun;69(1-2):63-75.

[3] Wagner B, Salman MD. Strategies for two-stage sampling designs for estimating herd-level prevalence. *Preventive Veterinary Medicine.* 2004 Dec;66(1-4):1-17.

[4] Cannon RM, Roe RT. Livestock disease surveys: a field manual for veterinarians. *Canberra: Australian Bureau of Animal Health*; 1982.

[5] Vose D. *Risk analysis: a quantitative guide.* 3rd ed. Chichester: Wiley; 2008.

[6] Krishnamoorthy K. *Handbook of statistical distributions with applications.* Boca Raton: Chapman & Hall/CRC; 2006.

[7] Rosskopf M, Staub E, Ackermann M. Comparison of 2 ELISA systems for detection of antibodies against IBR/IPV and against bovine leukemia-virus. *Schweiz Arch Tierheilkd.* 1994 Feb;136(2):58-67.

[8] Nuotio L, Neuvonen E, Hyytiainen M. Epidemiology and eradication of infectious bovine rhinotracheitis/infectious pustular vulvovaginitis (IBR/IPV) virus in Finland. *Acta Veterinaria Scandinavica.* 2007;49(3).

[9] Suzuki K, Marmo F, Guarino H. Monitoring in infectious disease status of dairy cattle herds in Uruguay: a random effects hierarchical model analysis In: Di Alberto P, Costa C, editors. *New research on livestock science and dairy farming.* New York: Nova Science Publishers; 2008. p. 228.

[10] Paisley LG, Tharaldsen J, Jarp J. A retrospective analysis of the infectious bovine rhinotracheitis (bovine herpes virus-1) surveillance program in Norway using Monte Carlo simulation models. *Preventive Veterinary Medicine*. 2001 Jul;50(1-2):109-25.

[11] Gall D, Nielsen K. Comparison of some methods for determining cutoff values for serological assays: A retrospective study using the fluorescence polarization assay. *Journal of Immunoassay & Immunochemistry.* 2001;22(2):85-98.

In: Agricultural Production
Editors: Felix C. Wager, pp. 93-103

ISBN: 978-1-61668-695-6

Chapter 4

SEROLOGICAL MONITORING OF SELECTED AVIAN PATHOGENS OF COMMERCIAL POULTRY FARMS IN URUGUAY

***K. Suzuki**[*1], **M. Petruccelli**[1], **G. Giossa**[1, 2],*
***G. Rodríguez**[1, 2], **G. Trenchi**[1, 2] **and H. Trenchi** [1, 2]*
[1]Universidad Nacional de La Plata, Buenos Aires, Argentina
[2]Universidad de la República, Montevideo, Uruguay

ABSTRACT

Poultry production is a growing industry in Uruguay, South America. *Mycoplasma gallisepticum* (MG) and *Mycoplasma synoviae* (MS) are the etiological agent of chronic respiratory disease in chickens and may be horizontally and vertically infected. These economically important diseases can influence commercial poultry operations in Uruguay. Chickens infected with MG and/or MS become lessening in egg productivity, hatchability and feed efficiency and grow in mortality, waste carcass and veterinary costs. Increasingly, these possible performance and production deprivations require the strict control of MG and MS amongst all the poultry sectors. The objective of this study was to estimate the seroprevalence of the two pathogens mentioned above in relation to Mycoplasmosis in commercial chickens in Uruguay. Sera randomly selected from flocks of Uruguay were analysed using a commercial enzyme-linked immunosorbent assay (ELISA). Test prevalence were adjusted for the imperfect test characteristics using the Rogan–Gladen estimator (deterministic and stochastic functions) and Bayesian inference. Overall, 0.4% and 0.9% of the tested 1861 chickens were classified as test-positive for antibodies against MG and MS, respectively. Considering the simulated ELISA specificities of 99.5% for MG and 99.2% for MS and in the absence of other reported diagnoses from the area, these results were therefore considered to be "false-positives". The ELISA in combination with testing sera and statistical methods for true prevalence

* Corresponding author: Tel./ fax: +54 221 425 3276; E–mail address: pvs@provetsur.net (K. Suzuki). PROVETSUR, Facultad de Ciencias Veterinarias, Universidad Nacional de La Plata, Avda. 60 y 118, La Plata, B1900AVW, Buenos Aires, Argentina

estimation provided a good survey tool at lower costs and time when compared to other approaches, and can be adapted to many other issues in animal health and production.

INTRODUCTION

Prevalence is a scale of animal disease frequency that concentrates on existing status rather than new events [1, 2]. Although incidence is of primary importance in etiological study, prevalence is usually used in the research of chronic animal diseases of insidious onset where incident cases are hard to define even if animal populations are monitored longitudinally. Prevalence is a dimensionless, unit-free value ranging from zero to one (zero to 100 if expressed as a percentage). Depending on the context, a researcher might be interested in prevalence of infection, infectious animals or disease. The authors restrict our presentation to infection prevalence hereinafter. Two types of prevalence are often estimated in animal-health studies: point and period prevalence. Point prevalence is the proportion of infected individual animals in a defined population (or more normally that have a condition of interest) at a given point in time; thus:

- Point prevalence =Number of infected individuals (e.g. chickens) / Total number of individuals at risk for infection in the population (e.g. flock).

Period prevalence is the proportion of infected individual animals in a defined population found over a defined time period (e.g. one year). For non-resolvable infections, point prevalence gives valid estimates of the frequency of individuals that have ever been infected and hence the authors do not further deal with period prevalence hereinafter.

Diagnostic tests are regularly used for prevalence studies and, preferably, true prevalence (TP) should be estimated from apparent prevalence (AP) by adjusting for test sensitivity (Se) and specificity (Sp). Lack of knowledge of or neglect for test errors (false positives and negatives) can lead to inappropriate sample size calculations for studies, misclassification of diseased and non-diseased states, and biased estimates of measures of effect in risk factor studies. All of these opposingly impact disease surveys, control and eradication programmes and, consequently, animal trade. Frequentist methods (which begin with a hypothesis test that proposes a specific value for an unknown parameter) have traditionally been used for animal-health prevalence surveys. However, in recent years, applications of Bayesian analytic methods (which are concerned with the consequences of modifying our previous beliefs as a result of receiving new data) for animal-health prevalence survey data have increased [3-9]. Bayesian statistical analysis of prevalence data is appealing as it formally integrates previously-collected prevalence data with expert-elicited information into current calculations [10]. In addition, when sample sizes for prevalence surveys are small, prevalence is low and results are based on imperfect diagnostic tests such as enzyme-linked immunosorbent assay (ELISA), it may be hard to satisfy large-sample normal approximations when frequentist confidence intervals (CIs) are constructed. In these circumstances, use of Bayesian approaches gives a practical alternative for data analysis.

Mycoplasma spp are most important bacterial pathogens that can cause respiratory problems in chickens and turkeys and/or be part of the chronic respiratory syndrome in chickens. Some of the primary causes of disease-associated loss in poultry production are related to infection with *Mycoplasma gallisepticum* (MG) and/or *Mycoplasma synoviae* (MS). Economic loss associated with Mycoplasma infection is usually the result of a complicated interaction between the *Mycoplasma* sp and co-infections with other bacteria, poor health management and/or poor environmental conditions. Studies have shown that over 80% of poultry operations worldwide are infected with MG and/or MS, making it one of the most prevailing and costly poultry diseases [11]. Even at low levels of infection, this chronic respiratory disease presents serious added costs to poultry operations through decreased feed efficiency, repeated medical treatments, poor chick quality and injurious vaccine reactions. The MG is linked with chronic respiratory disease in chickens. The symptoms normally seen are coryza, coughing, nasal exudates and respiratory rales. Economic losses due to MG infection comprise reduced egg production, poor eggshell quality, lowered hatchability of chicks and downgraded meat quality. The MS is a known pathogen connected with the development of synovitis and chronic respiratory disease in chickens. Clinical symptoms comprise joint swelling, coryza and respiratory rales. Economic losses due to MS infection include reduced egg production, lowered hatchability of chicks and downgraded meat quality [11, 12].

The objective of this study was to estimate the true prevalence of seropositive individual chickens against *Mycoplasma gallisepticum* and *Mycoplasma synoviae* using the Rogan-Gladen estimator [13] and Bayesian inference.

Materials and Methods

Study Area

Uruguay (conventional long form: Oriental Republic of Uruguay) is located in the south-eastern part of South America bordering the South Atlantic Ocean, between Argentina in the west and Brazil in the northeast. The population is currently estimated at 3.4 million people, of which some 3.1 million live in the capital city Montevideo, its surroundings and other urban areas. Ten percent of the total population engage in agricultural sector [14]. The climate in Uruguay is temperate with mean monthly temperature varying from 11 °C in winter and 24 °C in summer. Mean annual rainfall varies between 1000 mm in the south and 1600 mm in the north [15]. The total land area is 175,020 km^2 and 77% of the area is occupied by permanent pasture that is suitable for livestock production [14]. Uruguay has a poultry population of 14 million, a poultry meat production of 45,000 tonnes per year and a poultry egg production of 43,600 tonnes per year [14]. The south side of the country including the capital city Montevideo and Canelones Department had the concentration of chicken population (about 90% of the total), because of in-and-around the big market Montevideo [16].

Sample Collection

Seventeen flocks of broiler chickens greater than 35 days of age were kept under observation. Each study flock was randomly selected at different farms recruited from the capital city Montevideo, Canelones or Lavalleja (east of Canelones) Departments. Nothing has been reported on clinical symptoms of a MG/MS infection for at least a year. None of the chickens had been vaccinated against MG/MS prior to sampling. The required total sample size of 1533 from a chicken population of 14 million was sufficient to produce a 95% CI with a desired precision of ±2.5% when the estimated AP was 50%. The sample size in each of the farms was proportionally allocated (1% each of the total number of chickens at study farms) by the accessible financial, human and material means. The field investigation was conducted between October 2008 and April 2009, consisted of data collection through questionnaire interviews for each farm recruited, in conjunction with blood sample collections for each chicken (questionnaire results were not dealt with hereinafter).

Laboratory Examinations

Blood samples collected were used for diagnostic investigations. Sera were analysed using a commercial ELISA for the detection of antibody to MG and MS in chicken serum (FlockChek® *Mycoplasma gallisepticum* Antibody Test Kit and *Mycoplasma synoviae* Antibody Test Kit, IDEXX Laboratories, Westbrook, ME, USA). Positive and negative controls were included for each series of samples analysed. Absorbance was read on an ELISA reader at 650 nm. On the basis of the instruction manual of the ELISA kits, serum samples with sample to positive ratios greater than 0.5 (titres greater than 1076) were considered seropositive.

Data Analysis

Data were entered into a database using the Base in the OpenOffice.org software version 3.1.1 (Sun Microsystems, Santa Clara, CA, USA). Seroprevalence estimates based on the use of an imperfect test must be corrected to take account of test performance. Based on the published ELISA sensitivity (91.6%) and specificity (98.2%) values [17], estimated TP of antibodies among study chickens at each farm were calculated as follows:

(1) TP for each farm were derived from the AP using the Rogan-Gladen estimator [13] and information about the Se and Sp:

$$\text{TP} = (\text{AP} + \text{Sp} - 1) / (\text{Se} + \text{Sp} - 1)$$

The CI can be calculated from the variance estimate, given by:

$$\operatorname{var}(\hat{p}) = \frac{p(1-p)}{n(Se+Sp)^2}$$

where p is AP, and n is the number of chickens randomly selected.

The CI is:

$$\left(\hat{p} - \left(Z_{\frac{\alpha}{2}} \times \sqrt{\mathrm{var}(\hat{p})}\right), \hat{p} + \left(Z_{\frac{\alpha}{2}} \times \sqrt{\mathrm{var}(\hat{p})}\right)\right)$$

where $Z_{\alpha/2}$ is the $1 - \alpha/2$ percentile of the standard normal distribution (e.g. 1.96 for 95% CI).

For estimation of TP (deterministic function with 95% CI) above, Survey Toolbox software version 1.04 was used [18]. Estimated values for TP and its CI of less than zero were rounded up to zero by the within-function of the software.

(2) Consider estimation of the infection seroprevalence for a single flock where *y* chickens tested positive out of *n* chickens randomly selected. If the flock size (*N*) is much larger than *n*, then the sampling distribution of *y* is approximately binomial:

$$y|\pi, \mathrm{Se}, \mathrm{Sp} \sim \mathrm{Binomial}\ (n, \pi\mathrm{Se} + (1 - \pi)(1 - \mathrm{Sp}))$$

where π is the true seroprevalence of infection in the flock (TP) and Se and Sp are the sensitivity and specificity, respectively, of the diagnostic test applied to each chicken sampled. The authors modelled uncertainty about the Se and Sp of the diagnostic test using independent beta prior distributions [19]:

$$\mathrm{Se} \sim \mathrm{Beta}\ (d + 1, n - d + 1),$$

$$\mathrm{Sp} \sim \mathrm{Beta}\ (d + 1, n - d + 1)$$

where *d* is the number of desired (positive or negative) outcomes, and *n* is the number of samples tested per farm [19]. A beta distribution provides a flexible means of modelling uncertainty about parameters ranging from 0 to 1. A beta prior distribution also may be used for the infection seroprevalence.

$$\mathrm{TP} \sim \mathrm{Beta}\ (d_f + 1, n_f - d_f + 1)$$

where *d* is the number of desired (positive or negative) outcomes, *n* is the number of samples tested per farm, and the subscript '*f*' indicates a farm-specific value. For estimation of TP (stochastic function with 95% CI) above, the spreadsheet software Microsoft Excel 2007 (Microsoft Corporation, Redmond, WA, USA) in combination with the simulation add-in software @Risk for Windows version 5.0 Pro (Palisade Corporation, Newfield, NY, USA) with 10,000 iterations were used. Estimated values for TP and its CI of less than zero were rounded up to zero manually.

(3) A Bayesian model was used to derive posterior Bayesian estimates (denoted TP_B, Se_B and Sp_B) from prior distributions and the data from each farm in this study. The same prior beta distributions for Se_B and Sp_B as in the Rogan-Gladen approach mentioned above were used. A beta prior distribution also may be used for the infection seroprevalence, but this would imply that TP_B ($\pi = 0$) = 0 [20]. However, if the flock is truly not infected, then $\pi = 0$ and so we require a model that will allow for this. The infection seroprevalence using a mixture distribution was modelled:

$$\mathrm{TP} \sim \mathrm{Beta}\ (d_f + 1, n_f - d_f + 1) \text{ with probability } \tau$$

TP = 0 with probability $1 - \tau$

where d is the number of desired (positive or negative) outcomes, n is the number of samples tested per farm and the subscript 'f' indicates a farm-specific value, and τ is the probability that the flock is infected. With this mixture distribution, computation of the posterior probability that the flock is not infected is possible and this computation can be performed easily using WinBUGS software version 1.4.3 [21] under binomial-sampling schemes. A beta prior distribution also can be used for τ. Alternatively, τ can be set equal to an expert-elicited constant (τ_0). The Markov chain–Monte Carlo (MCMC) simulation was run for 12,000 iterations of which the first 2,000 iterations were discarded as 'burn-in'. The posterior means and 95% CI (also called Bayesian credible interval") were recorded for the farm-specific TP_B estimates and for posterior estimates of the test characteristics, Se_B and Sp_B.

RESULTS

The 1861 chickens studied accounted for about 1% of the study chicken population and 0.01% of the total chicken population in Uruguay. Of all, only one sample from the farm with the ID of 9, ranked 4th of AP against MG and 8th of AP against MS, had test-positive against both MG and MS. The proportion of all for test-negative against both the two diseases was 98.8%. Table 1 shows the estimated seroprevalence against MG among the study chickens categorised by farms. The numbers of chickens sampled between the 17 study farms were varied from 30 to 224. Five out of the 17 farms had AP of greater than 0%, between 0.4% and 2.7%. For the five farms, point estimates of TP by Rogan-Gladen estimators (deterministic as well as stochastic functions) were lesser than of AP. The 95% CI for each point estimate of TP by Rogan-Gladen (stochastic function) were consistently wider than by Rogan-Gladen (deterministic function). The meaning of value "< 0.1" in the column of Bayesian inference includes "greater than 0%". All the point estimates of TP by Bayesian inference were therefore greater than 0%. The AP of greater than 0% were adjusted lesser by Bayesian inference, while AP of equal to 0% were adjusted greater by Bayesian inference. All the point estimates of TP by both Rogan-Gladen methods were well within the corresponding Bayesian credible intervals. The Bayesian posterior sampling means for the Se_B and Sp_B, estimated from the study, were 91.0% (95% CI: 86.0-94.9%) and 99.5% (95% CI: 99.1-99.9%), respectively. Table 2 displays the estimated seroprevalence against MS among the study chickens categorised by farms. The numbers of chickens sampled between the 17 study farms were varied from 30 to 224 (same as Table 1). Nine out of the 17 farms had AP of greater than 0%, between 0.5% and 7.5%. For the nine farms, point estimates of TP by Rogan-Gladen estimators (deterministic as well as stochastic functions) were lesser than of AP. The 95% CI for each point estimate of TP by Rogan-Gladen (stochastic function) were consistently wider than by Rogan-Gladen (deterministic function). The meaning of value "< 0.1" in the column of Bayesian inference includes "greater than 0%" (same as Table 1). All the point estimates of TP by Bayesian inference were therefore greater than 0%. The AP of greater than 0% were adjusted lesser by Bayesian inference, while AP of equal to 0% were adjusted greater by Bayesian inference. All the point estimates of TP by both Rogan-Gladen methods were well within the corresponding Bayesian credible intervals. The Bayesian posterior sampling means for the Se_B and Sp_B, estimated from the study, were 91.0% (95% CI: 86.0-94.9%) and 99.2% (95% CI: 98.6-99.8%), respectively.

Table 1. Estimated seroprevalence against *Mycoplasma gallisepticum* among broilers in Uruguay.

Farm ID	No. of broilers sampled	Apparent seroprevalence (%)	True seroprevalence [%, (95% CI)] Rogan-Gladen (deterministic)		Rogan-Gladen (stochastic)		Bayesian inference	
1	91	2.2	0.4	(0-2.0)	0	(0-4.9)	0.8	(0-4.3)
2	89	1.1	0	(0-1.2)	0	(0-3.2)	0.5	(0-3.0)
3	79	0	0	(0-0)	0	(0-1.6)	0.2	(0-1.7)
4	97	0	0	(0-0)	0	(0-1.1)	0.2	(0-1.4)
5	100	0	0	(0-0)	0	(0-1.0)	0.2	(0-1.4)
6	100	0	0	(0-0)	0	(0-1.0)	0.2	(0-1.3)
7	100	0	0	(0-0)	0	(0-0.9)	0.2	(0-1.4)
8	113	2.7	1.0	(0-2.5)	0	(0-4.9)	1.0	(0-4.6)
9	111	0.9	0	(0-0.9)	0	(0-2.3)	0.4	(0-2.4)
10	119	0	0	(0-0)	0	(0-0.7)	0.1	(0-1.1)
11	200	0	0	(0-0)	0	(0-0)	<0.1	(0-0.7)
12	65	0	0	(0-0)	0	(0-2.4)	0.3	(0-2.0)
13	224	0.4	0	(0-0.5)	0	(0-0.6)	0.2	(0-1.1)
14	223	0	0	(0-0)	0	(0-0)	<0.1	(0-0.6)
15	30	0	0	(0-0)	0	(0-7.1)	0.6	(0-4.6)
16	80	0	0	(0-0)	0	(0-1.6)	0.2	(0-1.7)
17	40	0	0	(0-0)	0	(0-5.1)	0.5	(0-3.4)
Total	1861	0.4	0	(0-0.2)	0	(0-0)	<0.1	(0-0.5)

Table 2. Estimated seroprevalence against *Mycoplasma synoviae* among broilers in Uruguay.

Farm ID	No. of broilers sampled	Apparent seroprevalence (%)	True seroprevalence [%, (95% CI)] Rogan-Gladen (deterministic)		Rogan-Gladen (stochastic)		Bayesian inference	
1	91	1.1	0	(0-1.1)	0	(0-3.0)	0.5	(0-2.8)
2	89	1.1	0	(0-1.2)	0	(0-3.2)	0.5	(0-3.0)
3	79	1.3	0	(0-1.3)	0	(0-3.8)	0.6	(0-3.4)
4	97	0	0	(0-0)	0	(0-1.0)	0.2	(0-1.4)
5	100	0	0	(0-0)	0	(0-1.0)	0.2	(0-1.4)
6	100	0	0	(0-0)	0	(0-1.0)	0.2	(0-1.3)
7	100	0	0	(0-0)	0	(0-1.1)	0.2	(0-1.4)
8	113	0	0	(0-0)	0	(0-0.7)	0.2	(0-1.2)
9	111	0.9	0	(0-0.9)	0	(0-2.3)	0.4	(0-2.4)
10	119	0	0	(0-0)	0	(0-0.7)	0.1	(0-1.1)
11	200	0.5	0	(0-0.5)	0	(0-0.7)	0.2	(0-1.3)
12	65	0	0	(0-0)	0	(0-2.3)	0.3	(0-2.0)
13	224	0	0	(0-0)	0	(0-0)	<0.1	(0-0.6)
14	223	1.3	0	(0-0.8)	0	(0-1.8)	0.4	(0-2.2)
15	30	3.3	1.7	(0-5.1)	3.1	(0-13.0)	1.8	(0-9.3)
16	80	5.0	3.6	(1.0-6.1)	2.9	(0-9.4)	2.4	(0-8.3)
17	40	7.5	6.3	(2.0-10.6)	6.9	(0-16.9)	4.6	(0-14.0)
Total	1861	0.9	0	(0-0.2)	0	(0-0.3)	0.1	(0-1.0)

DISCUSSION

Detection of Mycoplasma infections can be limited by the sample size used during flock surveys. If the survey programme includes fewer than 30 samples per population, the risk of missing the detection of the first infected animal (first index case) is increased. Early detection of a recent infection is dependent on the frequency of sampling, sample size and the Se of the test. Also, there is not a serological test that has both 100% Se and 100% Sp available for *Mycoplasma* spp. To commence a periodical monitoring programme in a unknown situation, the necessary initial step should be to determine the prevalence of the *Mycoplasma* spp in each company or region [17]. Even the smallest sample size of 30 (the 15th farm) in this study met the expectations above. Therefore, all the AP estimated in this study can be used to know the situation in each farm.

The authors used both a Rogan–Gladen estimation (deterministic and stochastic functions) and Bayesian inference. The Rogan–Gladen approach has the advantage that it is widely known and also can be used as a simple deterministic function (entering fixed values for AP, Se and Sp) which is not needed any specific software for calculation. The Bayesian model approach is more complicated but relatively easily can be conducted in software WinBUGS which is freely available. Its advantage is to provide posterior distributions (estimates) for Se and Sp as well as TP, which can be useful to validate a commercial ELISA for detecting any agent in an unknown setting. However, knowledge and presumptions on the prior distributions, value range and starting values of the model inputs are required for implementing stochastic process.

The point estimates of TP against MG and MS by both Rogan-Gladen estimators (deterministic and stochastic) were lesser than of AP. This is because of including the concepts of Se and Sp (deterministic) and probabilistic distribution (stochastic) for adjustment, in comparison with the common approach obtaining AP. The 95% CI for each point estimate of TP against MG and MS by Rogan-Gladen estimator (stochastic function) were consistently wider than by Rogan-Gladen estimator (deterministic function). This is because including the concepts of modelling uncertainty (stochastic function) for adjustment, in comparison with the deterministic function. All the point estimates of TP against MG and MS by Bayesian inference were greater than 0%. Using Bayesian inference improves the disadvantage of Rogan-Gladen estimator that can yield negative results (less than 0) in the case of certain combinations of AP, Se and Sp.

The application of the Rogan–Gladen estimator is often controversial in some situations because TP can be greater than 1 or less than 0. In the context of MG/MS infection, only the former is of empirical importance and can occur in a small sample with low infection seroprevalence when the proportion of false-positive results is greater than the AP. The limitations of the Rogan–Gladen estimator are further exacerbated by the fact that when sample sizes are small and TP is low, the normal approximation used for calculation of CIs may not be justifiable.

A Bayesian analysis normally accounts for uncertainty in the values of TP, Se and Sp. Under binomial sampling, the distribution of the number of chickens testing seropositive is given by $y|\text{TP, Se, Sp} \sim \text{Binomial}(n, \text{AP})$, where $\text{AP} = \text{TP}*\text{Se} + (1 - \text{TP})(1 - \text{Sp})$. Inherent in the use of a beta prior for the seroprevalence is the certainty that $\text{TP} \neq 0$. If one wish to allow for the chance of zero infection prevalence, then one can use the mixture prior distribution:

TP ~ Beta (d_f + 1, n_f – d_f + 1) with probability τ and TP = 0 with probability=1 -τ. The parameter τ designates the possibility that the flock is infected, and in the single flock setting is typically set equal to an expert-elicited value, but can alternately be modelled using a beta distribution [20]. A mixture prior provides a more appropriate statistical model for low prevalence or disease freedom surveys. Models that do not allow for zero prevalence will be likely to generate estimates that are biased upwards. In addition to modelling TP, as previously mentioned, beta distributions are normally used as priors for Se and Sp [19, 20]. Considering the simulated ELISA specificities of 99.5% for MG and 99.2% for MS and in the absence of other reported diagnoses from the area, these results were therefore considered to be "false-positives".

CONCLUSION

Frequentist approaches to TP estimation are uncomplicated to use but suffer from severe statistical limitations when prevalence is low, sample sizes are small and uncertainty in Se and Sp is to be included into the precision of estimates. These limitations are simply improved in a Bayesian context and the authors recommend that researchers use Bayesian methods for prevalence estimation, wherever relevant. The ELISA in combination with testing sera and statistical methods for TP estimation provided a good prevalence study tool at lower costs and time when compared to other approaches, and can be adapted to many other issues in animal health and production.

REVIEWED BY

Dr Son Thi Thanh Dang, Department of Veterinary Disease Biology, Faculty of Life Sciences, University of Copenhagen, Denmark.

ACKNOWLEDGMENTS

This study was carried out as part of the project for the capacity development for improvement of livestock hygiene in the southern part of South America through regional cooperation [in Spanish: Proyecto de desarrollo profesional continuo para los veterinarios del Sur (PROVETSUR)], funded by the Japan International Cooperation Agency.

REFERENCES

[1] Thrusfield M. *Veterinary epidemiology*. 3rd ed. Oxford: Wiley-Blackwell; 2005.

[2] Dohoo I, Martin W, Stryhn H. *Veterinary epidemiologic research.* Prince Edward Island, Canada: AVC Inc.; 2003.

[3] Origlia J, Suzuki K, Castro L, Faccioli M, Silva M, Caballero J, et al. Bayesian mapping for infectious bronchitis virus risk in backyard chickens in Paraguay. *International Journal of Poultry Science.* 2009;8(8):740-5.

[4] Panei CJ, Suzuki K, Echeverría MG, Serena MS, Metz GE, González ET. Association of BoLA-DRB3.2 alleles with resistance and susceptibility to persistent lymphocytosis in BLV infected cattle in Argentina. *International Journal of Dairy Science.* 2009;4(3):123-8.

[5] Suzuki K, Caballero J, Álvarez F, Faccioli M, Goreti M, Herrero M, et al. Stochastic estimation of optimal age for infectious bursal disease vaccination in broilers in Paraguay. *Journal of Veterinary Epidemiology.* 2009;13(1):46-52.

[6] Suzuki K, Caballero J, Álvarez F, Faccioli M, Goreti M, Herrero M, et al. The use of linear mixed models to estimate optimal vaccination timing for infectious bursal disease in broilers in Paraguay. *International Journal of Poultry Science.* 2009;8(4):363-7.

[7] Suzuki K, Marmo F, Guarino H. *Monitoring in infectious disease status of dairy cattle herds in Uruguay: a random effects hierarchical model analysis.* In: Di Alberto P, Costa C, editors. New research on livestock science and dairy farming. New York: Nova Science Publishers; 2008. p. 228.

[8] Suzuki K, Origlia J, Álvarez F, Faccioli M, Silva M, Caballero J, et al. Relative risk estimation for Mycoplasma synoviae in backyard chickens in Paraguay. *International Journal of Poultry Science.* 2009;8(9):842-7.

[9] Suzuki K, Pecoraro MR, Gonzalez ET, Loza A, Perez M, Ruiz G, et al. Antibody seroprevalences against rabies in dogs vaccinated under field conditions in Bolivia. *Tropical Animal Health and Production.* 2008;40(8):607-13.

[10] Baadsgaard NP, Jogensen E. A Bayesian approach to the accuracy of clinical observations. *Preventive Veterinary Medicine.* 2003;59(4):189-206.

[11] Kleven SH. *Poultry: Mycoplasmosis.* In: Kahn CM, editor. The Merck veterinary manual, 9th edition. Whitehouse Station: Merck & Co.; 2005. p. 2242-6.

[12] Ley DH. Mycoplasma gallisepticum Infection. In: Saif YM, Barnes HJ, Glisson JR, Fadly AM, McDougald LR, Swayne DE, editors. *Diseases of poultry,* 11th edition. Ames: Iowa State University Press; 2003. p. 722-43.

[13] Rogan WJ, Gladen B. Estimating prevalence from results of a screening-test. *Am J Epidemiol.* 1978;107(1):71-6.

[14] FAO. FAOSTAT. 2009 [updated 2009; cited 26 October 2009]; Available from: *http://faostat.fao.org/site/291/default.aspx.*

[15] *FAO. FAOCLIM 2: world-wide agroclimatic data* (CD-ROM database). Rome; 2001.

[16] Guarino H. *Informe de la delegación uruguaya para la reunión de coordinación de PROVETSUR* (report submitted to the annual meeting of the project for the capacity development for improvement of livestock hygiene in the southern part of South America through regional cooperation). La Plata; 2007.

[17] IDEXX. Monitoring mycoplasmas in poultry farms (Animal health updates). 2006 [updated 2006; cited 26 October 2009]; Available from: *http://www.idexx.com/view/xhtml/en_us/livestock-poultry/newsletter/2006/200604.jsf.*

[18] Cameron AR. *Survey toolbox - a practical manual and software package for active surveillance of livestock diseases in developing countries.* Canberra: Australian Centre for International Agricultural Research; 1999.

[19] Vose D. *Risk analysis: a quantitative guide.* 3rd ed. Chichester: Wiley; 2008.

[20] Branscum AJ, Gardner IA, Johnson WO. Bayesian modeling of animal- and herd-level prevalences. *Preventive Veterinary Medicine*. 2004;66(1-4):101-12.

[21] Lunn DJ, Thomas A, Best N, Spiegelhalter D. WinBUGS - a Bayesian modelling framework: concepts, structure, and extensibility. *Statistics and Computing*. 2000;10(4):325-37.

In: Agricultural Production
Editors: Felix C. Wager, pp. 105-115
ISBN: 978-1-61668-695-6

Chapter 5

ASSESSING THE SEROPREVALENCE FOR ABORTIGENIC DISEASES OF OVINE HERDS BASED ON DIAGNOSTIC TESTS APPLIED TO INDIVIDUAL–ANIMALS IN URUGUAY

K. Suzuki*[*1], G. Travería[1], M. Cattáneo[1, 2], R. Puentes[1, 2], M. Martinicorena[1, 2], J. Moreno[1, 2], A. Furtado[1, 2], A. Freyre[1, 2], D. Satragno[1, 2], C. Acevedo[1, 2], R. Nuñez[1, 2] and J. Bermúdez[1, 2]

[1]Universidad Nacional de La Plata, Buenos Aires, Argentina
[2]Universidad de la República, Montevideo, Uruguay

ABSTRACT

The objective of this study was to estimate the true prevalence of seropositive individual sheep against *Chlamydophila abortus* and *Brucella ovis* using the Rogan–Gladen estimator in combination with Bayesian inference in Uruguay, South America. Ten ovine herds were kept under observation in 2008. Each study herd was randomly selected at different farms recruited from the Artigas (no. of the study herds = 1), Salto (no. of the study herds = 8) and Canelones (no. of the study herds = 1) Departments. The required total sample size of 1385 was determined by power analysis. Blood samples collected were used for diagnostic investigations. Sera were analysed using a commercial ELISA for the detection of antibody to *Chlamydophila abortus* and *Brucella ovis* in sheep serum. Adjusted outcomes were required for precise comparison of seroprevalence estimates. The overall antibody seroprevalence of *Chlamydophila abortus* and *Brucella ovis* would be extremely low. Considering the simulated ELISA specificities of 99.3% for *Chlamydophila abortus* and 97.0% for *Brucella ovis* and in the absence of other reported

* Corresponding author: Tel./ fax: +54 221 425 3276; E–mail address: pvs@provetsur.net (K. Suzuki). PROVETSUR, Facultad de Ciencias Veterinarias, Universidad Nacional de La Plata, Avda. 60 y 118, La Plata, B1900AVW, Buenos Aires, Argentina

diagnoses from the area, these results were therefore considered to be "false–positives". The methodology was useful for obtaining estimates of *Chlamydophila abortus* and *Brucella ovis* prevalence and for establishing prevalence distributions which could be used as input parameters in risk assessment and decision models.

INTRODUCTION

The quality of sheep farming in Uruguay is relatively high and the close interaction between producers, the extension services and the veterinary services acts a significant role in maintaining this condition. There is a growing concern that animal health problems may progressively obstruct the industry. This is especially the case with low reproductive performance which is normally recognised as the single most significant problem associated with the increasing numbers of sheep and concomitant increase in stocking rate [1].

Chlamydophila abortus (*Chlamydia psittaci* serotype 1) is the source of Enzootic Abortion of Ewes (EAE), which is distinguished by late term abortions, stillbirths, and weak lambs. Except for Australia and New Zealand, EAE exists worldwide and is most significant in intensively managed sheep. Abortions happen during the last 2–3 weeks of gestation regardless of when infection occurs, and the fetuses are fresh with minimal autolysis. Final diagnosis is by identification of *Chlamydophila abortus* by enzyme-linked immunosorbent assay (ELISA), fluorescent antibody staining, polymerase chain reaction (PCR), or isolation [2]. Ewes rarely abort more than once, but they last persistently infected and shed *Chlamydophila abortus* from their reproductive tract for 2–3 days before and after ovulation. Rams can be infected and transfer the organism venereally. Control consists of isolating all affected ewes and lambs and treating in–contact ewes with long-acting antibiotics such as oxytetracycline or oral tetracycline. *Chlamydophila abortus* bacterins are obtainable and are successful in reducing abortions. In parts of Europe, a modified live vaccine is obtainable for use. *Chlamydophila abortus* is zoonotic but human cases are infrequent. All have involved pregnant women, who developed life–threatening disease. Only in a few cases in which the fetus was delivered by cesarean section did the infant live. Pregnant women should not work with pregnant sheep, especially if abortions are occurring [3, 4].

The crucial importance of *Brucella ovis* is as a cause of epididymitis in rams, but it also causes late–term abortions, stillbirths, and birth of weak lambs. *Brucella ovis* sometimes causes abortion in sheep. *Brucella* abortions happen late in gestation, resulting in placentitis with edema and necrosis of the cotyledons and thickened, leathery intercotyledonary areas. Many fetuses aborted due to *Brucella ovis* are alive at the beginning of parturition, even though fetuses can be mummified or autolysed. Most fetuses aborted because of *Brucella ovis* are autolytic. Culture of the placenta, abomasal contents, and the dam's vaginal discharge are diagnostic. A vaccine for *Brucella ovis* is obtainable in some countries. *Brucella ovis* is non-zoonotic [3, 5].

Prevalence is a scale of animal disease frequency that concentrates on existing status rather than new events. Diagnostic tests are regularly used for prevalence studies and, preferably, true prevalence (TP) should be estimated from apparent prevalence (AP) by adjusting for test sensitivity (Se) and specificity (Sp) [6]. In recent years, applications of Bayesian analytic methods (which are concerned with the consequences of modifying our previous beliefs as a result of receiving new data) for animal-health prevalence survey data have increased. Use of Bayesian approaches gives a practical alternative for data analysis [7, 8].

The objective of this study was to estimate the TP of seropositive individual sheep against *Chlamydophila abortus* and *Brucella ovis* using the Rogan–Gladen estimator in combination with Bayesian inference.

MATERIALS AND METHODS

Study Area

Uruguay (conventional long form: Oriental Republic of Uruguay) is located in the south–eastern part of South America bordering the South Atlantic Ocean, between Argentina in the west and Brazil in the northeast. The population is currently estimated at 3.4 million people, of which some 3.1 million live in the capital city Montevideo, its surroundings and other urban areas. Ten percent of the total population engage in agricultural sector [9]. The climate in Uruguay is temperate with mean monthly temperature varying from 11 °C in winter and 24 °C in summer. Mean annual rainfall varies between 1000 mm in the south and 1600 mm in the north [10]. The total land area is 175,020 km^2 and 77% of the area is occupied by permanent pasture that is suitable for livestock production [11]. Beef cattle (11 million animals), dairy cattle (0.7 million animals) and sheep (10 million animals) compose the majority of the agricultural economic product. Livestock accounts for three quarters of agricultural GDP [12]. Because of the climatic characteristics, livestock production systems basically utilise open fields for direct grazing of natural and planted pastures [13]. Uruguay has a sheep meat production of 27,000 tonnes per year and a sheepskin production of 9,800 tonnes per year [9]. The north side of the country (Artigas and Salto Departments) had the concentration of sheep population (about 31% of the total population) [9, 11]. These two departments and Canelones Department north to the capital city Montevideo were selected as the study area.

Sample Collection

Ten herds of sheep were kept under observation. Each study herd was randomly selected at different farms recruited from the Artigas (no. of the study herds = 1), Salto (no. of the study herds = 8) and Canelones (no. of the study herds = 1) Departments. To our knowledge, nothing has been reported on seroprevalence of a *Chlamydophila abortus* / *Brucella ovis* infection for at least a year. None of the sheep had been vaccinated against *Chlamydophila abortus* / *Brucella ovis* prior to sampling. The required total sample size of 1385 from a sheep population of 10 million was sufficient to produce a 95% confidence interval with a desired precision of ±2.6% when the estimated AP was 50% [14]. The sample size in each of the farms was proportionally allocated (10% each of the total number of sheep at study farms) by the accessible financial, human and material means. The field investigation was conducted in 2008, consisted of data collection through questionnaire interviews for each farm recruited, in conjunction with blood sample collections for each sheep (questionnaire results were not dealt with hereinafter).

Laboratory Examinations

Blood samples collected were used for diagnostic investigations. Sera were analysed using a commercial ELISA for the detection of antibody to *Chlamydophila abortus* and *Brucella ovis* in sheep serum (CHEKIT® *Chlamydophila abortus* Antibody Test Kit and *Brucella ovis* Antibody Test Kit, IDEXX Laboratories, Liebefeld–Bern, Switzerland). Positive and negative controls were included for each series of samples analysed. Absorbance was read on an ELISA reader at 450 nm. On the basis of the instruction manual of the ELISA kits, each serum sample was determined as seropositive or seronegative.

Data Analysis

Data were entered into a database using the Base in the OpenOffice.org software version 3.1.1 (Sun Microsystems, Santa Clara, CA, USA). Meta–analysis deals with the problem of reaching consensus on a particular question, using evidence from multiple independent studies (e.g. 10 *Chlamydophila abortus* seroprevalence studies implemented in Uruguay) [15, 16]. In this example, meta–analysis is based on the belief that the 10 studies that were actually done can be treated as a random sample from the population of studies. The purpose of each of these 10 studies was to estimate the overall true seroprevalence, each of the 10 studies had to draw samples randomly from its study population. Consequently, a meta–analysis is essentially a sample of samples. Each study population has a true (but unobserved) seroprevalence, which it estimated from a sample of subjects. Seroprevalence estimates based on the use of an imperfect test, which is a nature of ELISA tests, must be corrected to take account of test performance. Based on the published ELISA sensitivity (73.3%) and specificity (96.3%) values for *Chlamydophila abortus* [17], and the published ELISA sensitivity (82.9%) and specificity (91.4%) values for *Brucella ovis* [18], estimated TP of antibodies among study sheep at each farm were calculated. TPs for each farm were derived from the AP using the Rogan–Gladen estimator [19] and information about the Se and Sp:

$$TP = (AP + Sp - 1) / (Se + Sp - 1)$$

A hierarchical model, which is a type of meta–analyses mentioned above, consists of one layer of sampling above another. Two hierarchical models were made for monitoring infectious disease status or *Chlamydophila abortus* and *Brucella ovis* antibody seroprevalence in Uruguay, respectively. Both models were fit using Bayesian methods which facilitate the estimation procedure for fitting complicated hierarchical models [20, 21]. A Bayesian model was used to derive posterior Bayesian estimates (denoted TP_B, Se_B and Sp_B, mentioned below) from prior distributions and the data from each farm in this study. Consider estimation of the infection seroprevalence for a single herd where y sheep tested positive out of n sheep randomly selected. If the herd size (N) is much larger than n, then the sampling distribution of y is approximately binomial:

$y|TP_B, Se_B, Sp_B \sim$ Binomial (n, AP)

where TP_B is the true seroprevalence of infection in the herd and Se_B and Sp_B are the sensitivity and specificity, respectively, of the diagnostic test applied to each sheep sampled, and $AP = TP_B*Se_B + (1 - TP_B)(1 - Sp_B)$. The authors modelled uncertainty about the Se_B and Sp_B of the diagnostic test using independent beta prior distributions [8]:

$Se_B \sim$ Beta $(d + 1, n - d + 1)$,

$Sp_B \sim$ Beta $(d + 1, n - d + 1)$

where d is the number of desired (positive or negative) outcomes, and n is the number of samples tested per farm. These values were decided by using the published Se and Sp values for *Chlamydophila abortus* and *Brucella ovis* mentioned above. A beta distribution provides a flexible means of modelling uncertainty about parameters ranging from 0 to 1 [22].

At the second level of the hierarchy, the model was to assume that percentage of test positive were alike in some way. This was equal to specifying a random effects model for the true seroprevalence probability p_i as follows. They were assumed to be drawn from a common Normal population distribution:

logit $(p_i) = b_i$

$b_i \sim$ Normal (μ, τ)

A standard non–informative prior is then specified for the population mean (logit) or probability of overall seroprevalence, μ, with an alternative non–informative prior considered for the random effects variance (a uniform prior on the standard deviation), because of the absence of strong prior information:

$\sigma \sim$ Uniform (0, 100)

$\tau = 1 / \sigma^2$

The true seroprevalence probability and associated 95% Bayesian Credible Intervals (BCIs) were computed via the Gibbs sampler, a Markov chain Monte Carlo (MCMC) technique, which was implemented using WinBUGS software [23]. The exponential of these true seroprevalence probabilities was taken to obtain overall seroprevalence estimates (Prev) and their 95% BCIs:

Prev = exp (μ) / (1 + exp (μ))

$\mu \sim$ Normal (0.0, 1.0E–6)

The built–in rank function in WinBUGS was used to distinguish studies which effect is in the extremes of this distribution. Results presented here were based on multiple runs of length 100,000 following a burn–in of 10,000 iterations to achieve convergence.

Table 1. Estimated seroprevalence against *Chlamydophila abortus* among sheep in Uruguay.

Farm ID	No. of sheep sampled	Apparent seroprevalence (%)	True seroprevalence (%)	95% Bayesian Credible Interval (%)	
				Lower	Upper
A	245	0.8	$2.3*10^{-4}$	0.0	0.0
B	54	0	$8.4*10^{-5}$	0.0	0.0
C	43	0	$1.5*10^{-4}$	0.0	0.0
D	98	0	$6.5*10^{-5}$	0.0	0.0
E	28	0	$3.2*10^{-4}$	0.0	0.0
F	368	0.3	$2.4*10^{-5}$	0.0	0.0
G	82	0	$1.1*10^{-4}$	0.0	0.0
H	82	2.4	$1.3*10^{-3}$	0.0	0.0
I	50	0	$3.5*10^{-5}$	0.0	0.0
J	308	0.6	$8.9*10^{-5}$	0.0	0.0

RESULTS

The 1358 (for testing *Chlamydophila abortus*) and 1365 (for testing *Brucella ovis*) sheep studied accounted for about 7.9% of the study sheep population and 0.01% of the total sheep population in Uruguay. Of all, none of the samples had test–positive against both *Chlamydophila abortus* and *Brucella ovis*. The proportion of all for test–negative against both the two diseases was 98.2%. Table 1 shows the estimated seroprevalence against *Chlamydophila abortus* among the study sheep categorised by farms. The numbers of sheep sampled between the 10 study farms were varied from 28 to 368. Four out of the 10 farms had AP of greater than 0%, between 0.3% and 2.4%. All the point estimates of TP by Bayesian inference were slightly greater than 0%. The APs of greater than 0% were adjusted lesser by Bayesian inference, while APs of equal to 0% were adjusted greater by Bayesian inference. The meaning of value "0.0" in the columns of BCIs includes "greater than 0%". The Bayesian posterior sampling means for the Se_B and Sp_B, estimated from the study, were 72.4% (95% BCI: 58.9–84.0%) and 99.3% (95% BCI: 98.8–99.7%), respectively. The overall true seroprevalence TP_B were $9.9*10^{-8}$% (95% BCI: 0.0–$2.2*10^{-42}$%). Table 2 displays the estimated seroprevalence against *Brucella ovis* among the study sheep categorised by farms. The numbers of sheep sampled between the 10 study farms were varied from 29 to 368 (similar to Table 1). Five out of the 10 farms had AP of greater than 0%, between 0.3% and 3.7%. All the point estimates of TP by Bayesian inference were slightly greater than 0%. The APs of greater than 0% were adjusted lesser by Bayesian inference, while APs of equal to 0% were adjusted greater by Bayesian inference. The meaning of value "0.0" in the columns of BCIs includes "greater than 0%". The Bayesian posterior sampling means for the Se_B and Sp_B, estimated from the study, were 82.2% (95% BCI: 73.3–89.5%) and 97.0% (95% BCI: 96.1–97.7%), respectively. The overall true seroprevalence TP_B were $1.4*10^{-9}$% (95% BCI: 0.0–$4.0*10^{-36}$%). Figure 1 and Figure 2 show the empirical distribution of the simulated rank of each component of the variable "Farm ID" for *Chlamydophila abortus* and *Brucella ovis* seroprevalence studies. These distributions demonstrate a spread of the rankings (high rank corresponds to low seroprevalence estimate, meaning high health status) and illustrate the considerable uncertainty associated with league tables (Table 1 and Table 2). These figures show that each farm has the almost same probability of 0.10 for any rank (between the 1st and

10th) regarding the seroprevalence against *Chlamydophila abortus* (Figure 1) and *Brucella ovis* (Figure 2) and report that there is little difference between farms regarding the seroprevalence against *Chlamydophila abortus* and *Brucella ovis*, respectively.

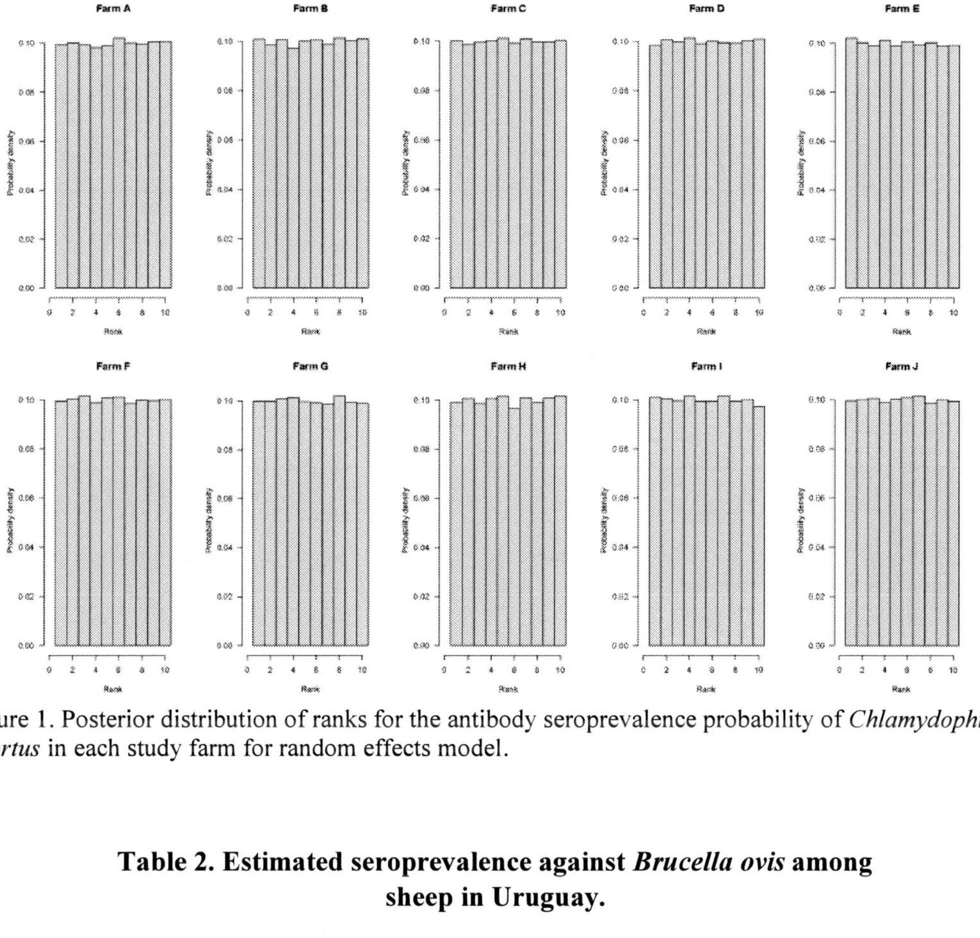

Figure 1. Posterior distribution of ranks for the antibody seroprevalence probability of *Chlamydophila abortus* in each study farm for random effects model.

Table 2. Estimated seroprevalence against *Brucella ovis* among sheep in Uruguay.

Farm ID	No. of sheep sampled	Apparent seroprevalence (%)	True seroprevalence (%)	95% Bayesian Credible Interval (%)	
				Lower	Upper
A	245	3.7	$1.6*10^{-4}$	0.0	0.0
B	56	0	$5.8*10^{-5}$	0.0	0.0
C	43	0	$2.3*10^{-4}$	0.0	0.0
D	100	2.0	$1.1*10^{-4}$	0.0	0.0
E	29	0	$2.6*10^{-4}$	0.0	0.0
F	368	1.1	$4.5*10^{-5}$	0.0	0.0
G	84	0	$8.3*10^{-5}$	0.0	0.0
H	82	1.2	$6.6*10^{-5}$	0.0	0.0
I	50	0	$5.8*10^{-5}$	0.0	0.0
J	308	0.3	$8.0*10^{-6}$	0.0	0.0

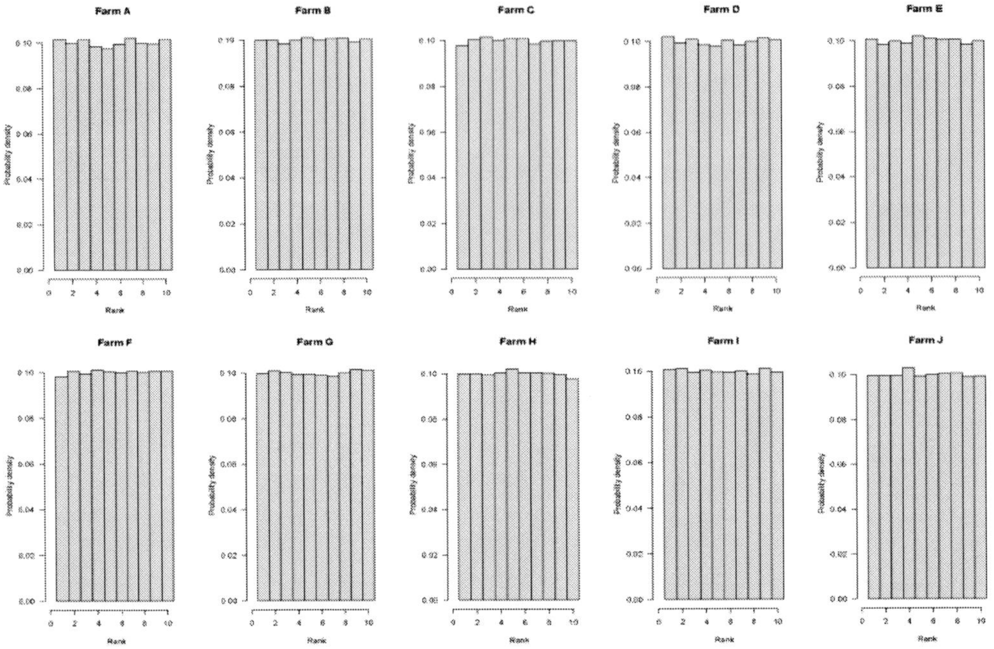

Figure 2. Posterior distribution of ranks for the antibody seroprevalence probability of *Brucella ovis* in each study farm for random effects model.

DISCUSSION

The present studies in Uruguay have investigated the seroprevalence estimates of *Chlamydophila abortus* and *Brucella ovis*. However, several factors differed between studies, including study area, study period and sample size. These variations between study designs make it difficult to draw generalisable conclusions regarding the prevalence of any particular infectious diseases. Meta–analysis strengthens the power of individual and relatively small studies by compiling results from independent studies [24]. The strengthened power leads to a higher precision of the estimates, by that means decreasing the variance and more accurately pointing out notable results.

Detection of *Chlamydophila abortus* and *Brucella ovis* infections can be limited by the sample size used during herd surveys. If the survey programme includes fewer than 30 samples per population, the risk of missing the detection of the first infected animal (first index case) is increased. Early detection of a recent infection is dependent on the frequency of sampling, sample size and the Se of the test. Also, there is not a serological test that has both 100% Se and 100% Sp available for *Chlamydophila abortus* and *Brucella ovis*. To commence a periodical monitoring programme in a unknown situation, the necessary initial step should be to determine the prevalence of the *Chlamydophila abortus* and *Brucella ovis* in each farm or region [25]. Except for the smallest sample size of 28 for *Chlamydophila abortus* and 29 for *Brucella ovis* (the Farm E), the sample sizes in this study met the expectations above. Therefore, almost all the APs estimated in this study can be used to know the situation in each farm.

The authors used a Rogan–Gladen estimation in combination with Bayesian inference. The Rogan–Gladen approach has the advantage that it is widely known and also can be used as a simple deterministic function (entering fixed values for AP, Se and Sp) which is not needed any specific software for calculation. The Bayesian model approach is more complicated but relatively easily can be conducted in software WinBUGS which is freely available. Its advantage is to provide posterior distributions (estimates) for Se and Sp as well as TP, which can be useful to validate a commercial ELISA for detecting any agent in an unknown setting. However, knowledge and presumptions on the prior distributions, value range and starting values of the model inputs are required for implementing stochastic process.

Adjusted outcomes are required for precise comparison of seroprevalence estimates. One of the aims of the present study was to illustrate how a hierarchical modelling approach permits the dependable estimation of the uncertainty corresponding an individual study's effect on outcome. The advantage of the approach taken in the study was that outcome data from all studies could be incorporated in one coherent inference framework, including small samples. The hierarchical model pooled data across all field investigations to calculate the prevalence and BCIs thus making comparative assessment more robust, and more reliable [23, 24].

Each league table of antibody seroprevalence of *Chlamydophila abortus* and *Brucella ovis* was made for rank ordering (Table 1 and Table 2). However, such presentations have been criticised as leading to a bogus focus on rank ordering, when it is known that the rank of prevalence is one of the most difficult quantities to estimate. The authors are able to quantify the variability in the prevalence ranks between studies. Advances in computing have meant that the MCMC methods are currently increasingly used to deal with wide variety of problems in statistics [23, 24]. The current models could easily be extended to a third level to compare prevalence estimates or infectious diseases status between neighbouring countries after adjustment for characteristics of study area and study period.

The overall antibody seroprevalence of *Chlamydophila abortus* ($9.9*10^{-8}$%) and *Brucella ovis* ($1.4*10^{-9}$%) would be extremely low. Freedom from such diseases gives several advantages, the most important of which is increased reproductivity. When demand is high, animals with a higher health status will also be worth more. In cases where disease prevalence is high and/or biosecurity level on a farm is low, it is feasible to manage animals by way of vaccination. This does not certainly rule out a farm's policy to finally produce an accredited disease–free herd.

All the point estimates of TP against *Chlamydophila abortus* and *Brucella ovis* by Bayesian inference were greater than 0%. Using Bayesian inference improves the disadvantage of Rogan–Gladen estimator that can yield negative results (less than 0) in the case of certain combinations of AP, Se and Sp. Considering the simulated ELISA specificities of 99.3% for *Chlamydophila abortus* and 97.0% for *Brucella ovis* and in the absence of other reported diagnoses from the area, these results were therefore considered to be "false–positives".

Conclusion

The methodology was useful for obtaining estimates of *Chlamydophila abortus* and *Brucella ovis* prevalence and for establishing prevalence distributions which could be used as input parameters in risk assessment and decision models. The analyses provide some guidelines for use when interpreting *Chlamydophila abortus* and *Brucella ovis* prevalence results and when comparing results from studies using different study designs (study area and study period particularly). The built–in rank function in WinBUGS was useful for investigating the considerable uncertainty associated with league tables regarding the seroprevalence.

ACKNOWLEDGMENTS

This study was carried out as part of the project for the capacity development for improvement of livestock hygiene in the southern part of South America through regional cooperation [in Spanish: Proyecto de desarrollo profesional continuo para los veterinarios del Sur (PROVETSUR)], funded by the Japan International Cooperation Agency.

REVIEWED BY

Dr Son Thi Thanh Dang, Department of Veterinary Disease Biology, Faculty of Life Sciences, University of Copenhagen, Denmark.

REFERENCES

[1] Bonino J, Cavestany D. *Aspectos de pérdidas reproductivas de origen infeccioso en ovinos* (in Spanish). Producción Ovina. 2005;17:69-76.

[2] Office International Des Epizooties. *Manual of diagnostic tests and vaccines for terrestrial animals*. 6th ed. Paris: Office International Des Epizooties; 2008.

[3] Nietfeld JC. *Abortion in large animals.* In: Kahn CM, editor. Merck veterinary manual. Whitehouse Station: Merck & Co., Inc.; 2005. p. 1097-108.

[4] Pospischil A. Enzootic abortion in ewes: A review of recent developments in diagnostics. *Small Ruminant Research.* 2006;62(1-2):113-5.

[5] Ficapal A, Jordana J, Blasco JM, Moriyon I. Diagnosis and epidemiology of *Brucella ovis* infection in rams. *Small Ruminant Research.* 1998;29(1):13-9.

[6] Martin SW, Meek AH, Willeberg P. *Veterinary epidemiology.* Ames: Iowa State University Press; 1987.

[7] Thrusfield M. *Veterinary epidemiology.* 3rd ed. Oxford: Wiley-Blackwell; 2005.

[8] Vose D. *Risk analysis: a quantitative guide*. 3rd ed. Chichester: Wiley; 2008.

[9] FAO. FAOSTAT. 2009 [updated 2009; cited 8 December 2009]; Available from: *http://faostat.fao.org/site/291/default.aspx.*

[10] *FAO. FAOCLIM 2: world-wide agroclimatic data* (CD-ROM database). Rome; 2001.

[11] FAO. Livestock sector brief - Uruguay. 2005 [updated 2005; cited 8 December 2009]; Available from: *www.fao.org/ag/againfo/resources/en/publications/sector_briefs/lsb_URY.pdf.*

[12] Ministerio de Ganadería Agricultura y Pesca (MGAP). Division Contralor de Semovientes (DI.CO.SE.). 2009 [updated 2009; cited 8 December 2009]; Available

from:
http://www.mgap.gub.uy/dgsg/DICOSE/dicose.htm.
[13] FAO. Grasslands of the world. 2005 [updated 2005; cited 8 December 2009]; Available from:
ftp://ftp.fao.org/docrep/fao/008/y8344e/y8344e00.pdf.
[14] Hintze J. *PASS* 2008. Kaysville, UT: NCSS; 2008.
[15] Dohoo I, Stryhn H, Sanchez J. Evaluation of underlying risk as a source of heterogeneity in meta-analyses: A simulation study of Bayesian and frequentist implementations of three models. *Preventive Veterinary Medicine.* 2007;81(1-3):38-55.
[16] Sanchez J, Dohoo I, Carrier J, DesCôeaux L. A meta-analysis of the milk-production response after anthelmintic treatment in naturally infected adult dairy cows. *Preventive Veterinary Medicine.* 2004;63(3-4):237-56.
[17] Vretou E, Radouani F, Psarrou E, Kritikos I, Xylouri E, Mangana O. Evaluation of two commercial assays for the detection of *Chlamydophila abortus* antibodies. *Veterinary Microbiology.* 2007;123(1-3):153-61.
[18] Alvarez JF, Veneros R, González O. Validación operacional de un ELISA comercial para Brucella ovis, Chile (in Spanish). *Archivos de Medicina Veterinaria.* 2007;39(3):275-9.
[19] Rogan WJ, Gladen B. Estimating prevalence from results of a screening-test. *American Journal of Epidemiology.* 1978;107(1):71-6.
[20] Papaspiliopoulos O, Roberts G. Stability of the Gibbs sampler for Bayesian hierarchical models. *Annals of Statistics.* 2008 Feb;36(1):95-117.
[21] Branscum AJ, Gardner IA, Johnson WO. Bayesian modeling of animal- and herd-level prevalences. *Preventive Veterinary Medicine.* 2004;66(1-4):101-12.
[22] Baadsgaard NP, Jogensen E. A Bayesian approach to the accuracy of clinical observations. *Preventive Veterinary Medicine.* 2003;59(4):189-206.
[23] Lunn DJ, Thomas A, Best N, Spiegelhalter D. WinBUGS - a Bayesian modelling framework: concepts, structure, and extensibility. *Statistics and Computing.* 2000 Oct;10(4):325-37.
[24] Dohoo I, Martin W, Stryhn H. *Veterinary epidemiologic research.* Prince Edward Island: AVC; 2003.
[25] Suzuki K, Marmo F, Guarino H. Monitoring in infectious disease status of dairy cattle herds in Uruguay: a random effects hierarchical model analysis. In: Di Alberto P, Costa C, editors. *New research on livestock science and dairy farming.* New York: Nova Science Publishers; 2009. p. 165-79.

In: Agricultural Production
Editors: Felix C. Wager, pp. 117-138

ISBN: 978-1-61668-695-6

Chapter 6

LAND TITLING AND AGRICULTURAL DEVELOPMENT

David S. Jones
University of Brunei, Jalan Tunkgu Link, Brunei

ABSTRACT

The chapter evaluates the effects of land titling on the output and incomes of farmers in developing countries, especially focusing on the small farm sector. In response to the large numbers of small farmers who have lacked formal title and therefore secure tenure, a number of governments in developing countries have in recent years implemented comprehensive titling programs. The chapter will consider a range of such programs and the evidence of the benefits they have brought to the small farm sector in terms of investment, credit access and land markets. The chapter will as well examine factors which have restricted the impact of titling, including lack of capital access in the small farm sector, inadequate extension services, market outlets and insurance cover, and either limited or excessive access to credit. Also considered are issues and challenges involved in titling communal tribal lands, and how these challenges can be met. In summary, the chapter will contend that land titling makes an important contribution to increasing farm output, but its beneficial impact is contingent upon the existence of other factors that must be addressed in conjunction with the titling program itself.

INTRODUCTION

Land titling (otherwise known as title registration) is the means by which the ownership interest of a property is legally recognized and officially recorded. As such it provides farmers with security of tenure, and is an important factor in promoting agricultural development by encouraging long term investment in farm enterprises, and by facilitating the transfer of land to more productive farmers through the creation of a land market. It also increases farmers' access to credit, which enables them to buy land and finance investment.

In developed countries, nearly all owner farmers have title to their land. By contrast, in many developing countries, a large majority of them, mainly in the small farm sector, until recently lacked title, leaving them without out legal recognition of their ownership interest and thus lacking security of tenure. This in turn has curtailed investment in farm enterprises and thus impeded improvements in output and living standards. In view of the acknowledged benefits of land titling, many of the governments of these countries in recent years have been prompted, often with the assistance of international donor organizations, to implement comprehensive titling programs, chiefly for the benefit of small farmers.

After discussing the features, processes and consequences of land titling, the chapter will examine how titling programs have been recently undertaken in the farming sector in a range of developing countries. It will then consider the beneficial impact of such programs on agricultural investment, credit acquisition, and the land market in the small farm sector. Also considered are the constraints which limit the positive impact of land titling. Here, attention will be given to the absence of other requirements essential to increasing agricultural output in developing countries, which thus undermines the impact of land titling. Lastly the chapter will consider the challenges and dilemmas that arise in the titling of communal tribal land in developing countries, and how these may be overcome by combining elements of both collective and individual titling.

Nature of Title Registration

Content of Title Register

Land title registration involves identifying a holding or parcel of land (sometimes called a lot) and determining the person(s) (or organization) with an ownership interest (or title) in it, which are then recorded in a land register. The details entered include the serial number assigned to the holding, and its location and boundaries, which are marked on a map. Also specified is the name of the person(s) with an ownership interest. It is usually required that the nature of the interest be specified. Not only freehold title may be registered but other forms of interest such as a life estate, long term leases (e.g. 50 year or 99 year leases), or long term or perpetual right of use. These other types of interest may considered as a form of limited ownership or derivative title under the superior interest of the absolute or freehold owner. In return, the holder of the title is given a certificate to prove his interest [Binns and Dale, 1995]. .

Ideally the entry in the land register indicates, as well, any easements and encumbrances attached to the land. Easements refer to the right given to others to enter and use the land for a specific purpose, such as a usufruct or profit à prendre right (e.g. harvesting fruit from trees on the land, or grazing the pasture), or the right to cross the land for the purposes of reaching a road, a source of water such as river or borehole, or another parcel of land (right of way). Encumbrances are burdens or liabilities that arise from financial claims on the land such as a mortgage, a right to a share of the income from the land, or a right to exact a rent. Also recorded in the land register may be specifications on how the parcel of land should be used, in line with lease covenants or zoning requirements [Binns and Dale, 1995].

The maps and the entry details relating to the holding may be confined to a single document in the land registry, or may be kept as separate documents in a folio. Alternatively and increasingly in many countries, the maps and information relating to the titling of a property are entered into an on-line title registration data base maintained by the registration authority.

Once title has been registered, based on the Torrens principle, the title holder has an indefeasible claim over the land that cannot be challenged or disputed. No further investigation is usually necessary to prove his/her interest, and no counter claim can be thus entertained. In that event, the title holder has a full legal safeguard of his/her interest, and this certainty is essential when the land is subsequently conveyed to another (Zevenbergen, 1998; Binns and Dale, 1995, 2-5; McEwan, 2001]. However, when the registration of a land holding is first undertaken, it may not be possible to prove that the title is beyond challenge both as to boundaries and title holder. In such cases, presumptive title (sometimes called qualified title, provisional or possessory title) is initially awarded. Although the title holder is presumed to be the legal and rightful owner of the interest, and the boundaries correct, these can be challenged within a period of time, and if the counter claim is upheld then the presumptive title is vitiated. If the period elapses without any counter claim, the presumptive title is converted into a full or absolute title [Binns and Dale, 1995].

Process of Title Registration

The process of title registration involves a number of stages. At the outset, a landholder submits a claim for title registration, or a locality is earmarked for en bloc registration of the land holdings in it. Existing cadastral maps and records (a map and record of land parcels) are consulted, or a new cadastral map for the locality is drawn up [McEwan, 2001]. This is normally followed by a survey of each land parcel which is the subject of the title claim, to determine its location, area, boundaries, usage and other features. At the same time, any documentary evidence submitted by the claimant(s) in support of the claim must be vetted. It is usual for on-site interviews to be conducted with the claimant, his/her neighbors, community leaders and other interested parties, to further assess the merits of the claim, secure agreement on boundaries, and clarify easement rights and encumbrances. This together with the survey of the holding may be referred to as adjudication (to be discussed below), and may be essential if documentary evidence is lacking [McEwan, 2001]. Following the survey and adjudication, a new cadastral map specifying the property of the claimant may be drafted. As happens, there may be discrepancies in the information gathered which must be resolved, e.g. the boundaries of the parcel of land in the existing cadastral map do not tally with the boundaries indicted by the claimant [McEwan, 2001]. Following this, a report is submitted to the registration authority to be vetted at different levels, and in some cases referred to other agencies for clearance. If there is nothing amiss, and after all matters have been resolved, formal title is then granted and recorded by the registrar or examiner of titles and a certificate of title issued to the claimant(s) [Zevenbergen, 1998]. In some countries, the granting of title is subject to the final approval and signature of the head of state.

In certain countries, land titling programs have been facilitated by the existence of other cadastral records. These may include records that certify land use and occupation, registers of deeds of conveyance, and land valuation records by which land taxes are levied. In Thailand, for example, for thirty years prior to its land titling program, a limited program of registration was undertaken which recognized and recorded temporary possession and right of usage of holdings. Those whose holdings were so recorded were awarded certificates of temporary occupation and certificates of land utilization. Although falling short of guaranteeing full ownership title, possession of such certificates is taken into account in deciding to award such title during the subsequent land titling program. Indeed, for certain types of certificate, little else apart from an on-site farm survey was required in order to be granted a full title [Slaats, 1999].

However, in most developing countries pre-title cadastral records, where they have existed, have been incomplete, inaccurate or have become out-dated. This has been compounded by the fact that many small farm households lack formal documents to support their claim to ownership such as tax receipts or deeds of transfer. In the absence of official records and personal documents, title is normally awarded on the basis of evidence of long-term and uncontested occupation gathered through the survey and adjudication process. In Cambodia, the land titling program, which began in 2001, has had to rely largely on field work comprising on-site surveys and adjudication as most of the cadastral records and personal documents were lost or destroyed during the Khmer Rouge regime and the ensuing civil war. If it is shown through surveys and adjudication interviews that the claimant has farmed the land for five years, during which time occupation was 'peaceful and uncontested', then title can be awarded [Ballard, 2006; Permanent Committee on GIS Infrastructure for Asia and the Pacific (PCGIAP) and Fédération Internationale des Géomètres (FIG), 2008].

Approaches to Title Registration

There are two approaches to land title registration. One is sporadic or ad hoc registration in which the occupant or user of a holding of land requests that his/her claim to title be recognized and his/her title be then registered. In support of the application, a survey of the land would be undertaken, with the claimant submitting relevant documentary evidence to support the claim, coupled if necessary by verbal testimony from all interested parties. The other approach is systematic registration. Here the land registration is initiated by the land registration authority (not by the landholder or land user), as part of a concerted land titling program. The registration officials undertake a simultaneous titling of many land parcels across a community, going from household to household, conducting surveys and adjudications for each holding. After the necessary vetting and endorsement by the registration authority and other relevant agencies (sometimes a lengthy process), the land holders are then awarded their individual titles at the same time [Holstein, 1996; World Bank (WB), 2003a; WB, 2003b].

Recent Land Titling Programs

In response to the fact that so much land was not registered, over the last 25 years, a number of developing countries have implemented systematic land titling programs. The way titling is undertaken has depended to an important degree upon the predominant system of land tenure, as discussed below.

Land Titling of Independent Farmers with Informal Ownership

A good deal of systematic land titling has occurred in countries where the bulk of the farming population are small independent peasant farmers cultivating their own land, whose ownership has been informal and thus not officially recognized. The point of the land titling program in such cases is to legally confirm the de facto ownership through adjudication, surveys, and where possible examination of documents supporting the claim to ownership.

One such program, which has been labeled as model for other countries, is that undertaken in Thailand. By 1980, under sporadic titling, only 12 per cent of agricultural properties had been titled. In 1985 a systematic land titling program began, assisted by the World Bank, and implemented by the Department of Lands in the Ministry of the Interior through 76 provincial land offices, 272 branch offices of the provincial land offices, and 758 district land offices. The program was divided into four phases and facilitated by the use of modern mapping and surveying equipment, use of satellite data, and the computerization of maps, survey data and cadastral records, as well as a large pool of well-trained professional personnel. It drew upon comprehensive cadastral records and cadastral and base maps already in existence. However, much field work was still undertaken in terms of on-site surveys and adjudication [Holstein, 1996; WB, 2003b].

Indonesia has adopted a similar program of granting independent peasant farmers full legal title. Title registration was initially authorized under article 19 of the Basic Agrarian Law of 1960. However, only a fraction of farm holdings were registered in the years following (an estimated 7 per cent of all rural properties were registered, and very few holdings of small farmers) [Slaats, 1999]. However, in 1995, a major reform of land registration was implemented with the introduction of the World Bank assisted Land Administration Project (LAP). Its remit was to undertake systematic registration coupled with mapping and surveying across different communities, with a view to giving landholders full ownership title (*hak milik*). The agency responsible for implementing the registration program was the National Land Agency (BPN).

Under LAP, teams of adjudicators and surveyors visited selected areas, going from house to house gathering verbal evidence of the occupation and use of the land. The household together with its neighbors were left to mark out the boundaries. If agreement was reached, then a report could be submitted recommending that the land be registered and the household given a title certificate. If agreement was not reached, the land remained unregistered. The title registration includes not only the name of the owner, location and boundaries of the land, but also encumbrances and usage restrictions attaching to the land [Heryani and Grant, 2004; Slaats, 1999; PCGIAP and FIG, 2008]. In keeping with the decentralization of public administration in Indonesia in 2001, much of the day to day work of registration has been passed from the BNP to District Authorities, although it still retains responsibilities for monitoring and checking land registration [Heryani and Grant, 2004].

Land Titling of Tenants and Workers on Large Estates

In some countries, land titling has been conducted in conjunction with or subsequent to agrarian reform by which large estates have been broken up and the land allocated to the estate tenants and landless workers. This has happened in the Philippines, where the bulk of the farming population were tenants (mainly sharecroppers) of large estates or latifundia. Although land registration began at the beginning of the last century, only a small fraction of the properties were registered, these being mainly large estates and plantations. A major impetus was given to land title registration by the Comprehensive Agrarian Reform Program (CARP) introduced in 1988. The tenants and landless groups who became owners of land or were allocated land have been given ownership certificates and eventually titles. Also under CARP, work is in progress to award land titles to small independent peasant farmers, outside the large estates, informally occupying holdings on public lands. When title registration is completed under CARP, the area of land on which holdings have been registered will comprise 6.8 million hectares [Quizon, 1999; Guardian, 2003]. However, the delays in transferring and distributing land, as well as protracted administrative procedures, have slowed the process of title registration.

In Bolivia too land titling has been mainly undertaken in the context of agrarian reform. The key measure which transformed land tenure was the 1953 Land Reform, by which the small tenants and landless workers (peons) received land through the acquisition and division of the large scale estates or hacienda. Little more that 30 per cent of the recipients of the land were awarded so-called agrarian reform titles, leaving most of the rest without proper legal title, who thus remained informal owners. In addition, those titles which were issued were often inaccurate in terms of boundaries and holding description, and rarely kept up to date in terms of land transfers, change of ownership, and alteration of holding boundaries. In view of this, a new program of systematic titling began in 1996, under the newly established National Institute for Agrarian Reform (INRA). The aim is both to issue titles to existing formal owners (those with old and usually outdated agrarian reform titles), and for the first time to informal owners (who can provide verbal or other documentary evidence to support a title claim) [De Mey and Bertón, 2004]. In the Cochabamba Tropics in the eastern part of Bolivia, significant progress has been made in new titling program, resulting so far in the grant of titles to over 70 per cent of the farm households [Chemonics International Inc., 2007].

Land Titling and the Break Up of Collective Farms

In communist states (and in states whose land reforms have been based on socialist principles), land titling has been conducted in conjunction with the break up of the collective farms under market reform policies. Such farms had been a mainstay of the socialist system and their dismantlement marked the beginning of the recognition of private property rights.

In Vietnam, a land titling program has been closely linked to the break up of collective farms under market reforms introduced by the Vietnamese government, known as the *Doi Moi* policy. The 1988 Land Law and the key implementing directive, Resolution 10, authorized the full scale division of the collective farms into individually worked parcels, vesting the assignees with land use rights over their parcels for 10-15 years [Iyer and Do, 2008; Ravallion and de Walle, 2003]. Under the Land Law of 1993 further amended by the

Land Law of 2003, these land use rights were formally converted to formal land use titles and the term of tenure was raised to 20 years for annual crop land and 50 years for perennial crop land. The land use rights could also be transferred, exchanged, bequeathed, rented and mortgaged. In effect, this amounted to granting the peasant population leasehold ownership. To ensure a reasonably equal distribution of land, limits were set on the amount of land allocated, which depended on the number of labor units in the family and the type of farming undertaken [Government of the Socialist Republic of Vietnam (GSRV) 1993; GSRV, 2003; Ngo, 1993; Iyer and Do, 2008]. Land use titles on the same terms were also accorded to peasant households in the south east and the Mekong Delta, who had not been assimilated into collective farms, and therefore continued to cultivate their own parcels of land. They were allowed to retain the parcel of land they had cultivated, but their informal ownership of the land was replaced by formal right of land use [Ha, 2007; Iyer and Do, 2008; Ravallion and de Walle, 2003]. Following a process of surveying and adjudication, households who had been allocated a land use right were issued with a land use right certificate (sometime called a Land Tenure Certificate or Red Book). The registration and Red Book, as well as specifying the area of the land holding, and its location, stipulated how it was to be used, and the period of tenure [Ha, 2007; Iyer and Do, 2008].

In Laos too following the commitment made in the 1980's to introduce market based reforms (which included the break up of the collective farms), two major land title registration programs began in the 1990's. One of them known as the Land Titling Project, began in 1997, allocated only land use rights to households and organizations, As in Vietnam, this is similar to leasehold ownership title, since it confers upon the household a full range of land rights, viz. to use, develop, sell, sub-lease, bequeath and mortgage the land on an indefinite basis.

Although not a communist state, Peru in the late 1960's and early 1970's implemented radical agrarian reforms along socialist lines. This resulted in the expropriation of the owners of large sugar plantations and haciendas, but the hacienda tenants and plantation workers, instead of being allocated individual land rights, became members of Agricultural Production Cooperatives. Some of the land of the cooperatives was worked on a collective basis, whilst the other portion was farmed collectively [Antle et al, 2003]. In the 1980's the majority of these were dismantled, with each household working in the cooperative allocated an individual holding but without formal title. In the 1990's a systematic program of lad titling began to convert the allotted land into registered holdings with the occupiers enjoying a full range of ownership rights [Antle et al, 2003].

Land Titling of Tribal Communities Under Communal Tenure

Land titling in certain developing countries has also been applied to lands which were or are occupied under so-called communal tenure by traditional tribal communities (variously called indigenous, ethnic or aboriginal communities). They live and farm in a range of habitats including desert areas, grasslands, forests, and upland areas, often in remotely situated regions. Under tribal communal tenure, land belongs to the community (which could be the tribe as a whole, or a sub-tribe, clan or lineage group within the tribe) and not to the individual. The occupation and use of the land is determined or sanctioned by the traditional leadership of the community and governed by customary rules which have evolved over generations. The right to use land may be inheritable but cannot be transferred to another person outside the community [Ezigbalike et al, 1996].

Various traditional systems of farming are associated with tribal land tenure and land use. One which is prevalent throughout savannah regions of Africa, is sedentary cultivation and common grazing. Tribal households cultivate permanent plots of land under the sanction of the tribal authorities, with ownership vested in the tribe itself. In addition, they commonly graze livestock on an open pasture reserve, in which grazing rights (the number of livestock each household is permitted and the section of pasture it can use) also determined by the tribal authority. In more arid regions of Africa and also in the Middle East and Central Asia, tribal groups engage in nomadic herding or pastoralism. They move their livestock across extensive reaches of territory throughout the year. The movement is dictated by the availability of water resources as determined by rainfall or boreholes, and in more mountainous regions seasonal changes in temperature. The pattern is often cyclical with the herders returning to the original pasture at the end of the year [Suttee et al, 2005].

A further type of traditional farming system found in tropical forest regions is shifting cultivation (commonly referred to as swidden, in Latin America as roza or roca, and in parts of India as jhum). Each tribal household is apportioned the right to use a parcel of land. After a period of cultivation, usually 1-4 years, the land is abandoned, necessitated by the depletion of soil fertility. In its place, the household moves to another piece of ground, which it then clears and cultivates, sometimes outside the immediate vicinity of the nucleated settlement in which most of the lineage group live. In due course, this land is also abandoned and so the process is continued. The abandoned land is often left as fallow so as to regenerate the soil, and then in time - perhaps after 8 to 15 years - re-cultivated as part of a rotational pattern. In addition, portions of forested land may be set aside to be used in common for hunting, gathering, and fishing. In some cases, as in Nepal, livestock are integral to shifting cultivation with the fallow land used for grazing [Aryal and Kerkhoff, 2008; Cleary and Eaton, 1996].

Land titling of communal tribal land has been particularly controversial and subject to much public debate. Many governments have resisted titling of communal tribal land, examples being those of Thailand and Nepal. Other governments have been more flexible and allowed titling in one form or another. For the latter, there are different options in determining how tribal land should be titled.

One option is to grant title on a collective basis to the community as a whole which occupies the lands, instead of awarding title to the individual households. In such cases, the community itself becomes the legal proprietor and is accorded the status of a legally recognized and registered association. The Philippines has adopted this approach,. Under the Indigenous People's Rights Act (IPRA) of 1997 collective ownership may be vested in so-called indigenous cultural communities with respect to their ancestral domains, usually found in the forested and upland areas. The legislation affirms the concept that such domains are 'private but community property which belongs to all generations' of the indigenous community, and 'cannot be sold, disposed or destroyed'. It does, however, allow for land to be transferred within the community. To further underline their protected ownership status, the Act asserts the right of an indigenous community 'to stay in the territory and not be removed therefrom', and 'to develop, control and use lands and territories traditionally occupied, owned or used' [Congress of the Philippines, 1997; Llanto and Ballesteros, 2003]. Cambodia too has recognized collective tenure of indigenous tribal communities living mainly in the northern and northeastern parts of the country, under the Land Law of 2002.

Such communities can now be treated as legal entities for the purpose of collective title registration [Guttal, 2006; Ballard, 2006]. In Ghana, under the 1992 constitution, title for much of the communal or tribal land was vested collectively in the tribe or clan, in the name of the paramount chief, clan head, or tribal council of elders. As title holders, their responsibility was to allocate to each household in the tribe or clan, rights to use parcels of land for residence and cultivation [Knox, 1998a].

Another option in titling communal tribal land is to divide part of the land into individual holdings, and allocate to each household a particular holding (which may consist of more than one plot of land). The allocation is accompanied or followed by the award of a land title. In Vietnam part of the lands occupied under communal tenure by indigenous tribal communities living in the northern and central mountainous areas of the country, have been divided in this way . The lands were allotted to the households of the tribal communities who then receive title to the land in the form of land use certificate, as mentioned above. The stipulation is that the households farm their land on a settled basis, so abandoning communal tenure and shifting cultivation. Up to 2007, about 70 per cent of the cultivated land of the communes where indigenous tribal communities are the predominant population, is now titled [General Statistics Office of Vietnam (GSOV), 2006]. In the 1990's the government of Laos followed a similar arrangement, dividing lands held under communal tenure by tribal groups, with each of the households allocated an individual holding under the requirement that the land is cultivated permanently and shifting cultivation is abandoned. However, the holding allocated normally consists of three small plots, intended to enable households to practice a short fallow rotation, with a portion of land still retained as a forest reserve that can be used on a communal basis. This in part mitigates the impact of being compelled to terminate conventional long fallow shifting cultivation. Each household is given at first a temporary certificate of land use, which after a number of years, can be converted to a full land use title certificate [Fujita and Phanvilay, 2004; Yokoyama et al, 2006; Evrard, 2004; Soulivanh et al, 2004].

In Botswana, a policy was adopted after independence which combines elements of the above two options, entailing both collective and individual title of tribal land. Here the main land use pattern of the predominant tribal group, the Tswana, is based on sedentary cultivation of individual holdings and common grazing on open ranges. Under the Tribal Land Act of 1968 ownership of the land was formally and collectively vested in each of the tribes of the Tswana group. The title was held by the land board of each tribe set up as the tribal authority to allocate land and regulate its use. Under the umbrella of collective tribal ownership, households continued to cultivate their individual holdings but were given formal land use rights to do so, specified in a Certificate of Customary Land Grant, which were to be enjoyed in perpetuity. These rights are inheritable, mortgagable, and to a degree saleable, so in effect constituting a form of leasehold ownership. Households also continued to use pasture reserves owned by the tribe for the common grazing of their livestock [Knox, 1998b]. In addition, under the Tribal Grazing Land Policy (TGLP) introduced in the 1970's (superseded by National Policy on Agricultural Development in 1991), specific areas of the tribal pasture (usually those which had been under-utilized) were divided into extensive commercial ranches and allocated to elite members of the tribe, often owners of boreholes who grazed large herds of cattle. The ranches were registered under a renewable lease title of 50 years but with the freehold ownership still residing with the tribe [Knox, 1998b; Frimpong, 1995].

Just over 419 ranches were created under the TGLP averaging 6,400 hectares, with just over two thirds managed by syndicates [Government of Botswana (GOB), 2007; Frimpong, 1995].

The Benefits of Land Titling in Agriculture

Encouraging Long-Term Investment

In developing countries, as elsewhere, increases in agricultural output depend upon major investment of fixed capital, the returns from which may only be realized over several years. Such investment includes the acquisition of vehicles, farm machinery and other implements, the use of high quality seeds, planting of saplings for tree crops, the installation of drainage and irrigation systems, and the application of a whole range of chemicals essential to improving the fertility of the soil. Also necessary are buildings where vehicles, equipment, tools, and seeds may be kept and after the harvest where crops may be stored.

Land titling is considered important in encouraging fixed capital investment in farming. With the security of tenure guaranteed by legal title, farmers have full confidence that they will be able to reap the long-term benefits of such investment. Without that guarantee, farmers have little legal protection against dispossession, encroachments, or rival claims of ownership. They must then to live with the uncertainty whether the long term benefits from fixed assets in which they have invested can be realized. With such uncertainty, farmers without title may be reluctant to undertake such investments. The connection between land titling and investment was strongly asserted in a World Bank report on land policy in 2003, and has also been reflected in the studies of other writers [WB, 2003a; Feder et al, 1988; Besley, 1995; Burns, 2006].

The positive impact of land titling on investment in farm enterprises and higher output per acre is evident in Vietnam. In the three regions of Vietnam where output per hectare was high, exceeding US$1,500, more than 80 per cent of the land was titled. In three of the five regions where output per acre was below this figure, the proportion of land which was titled was much lower, though in the other two regions of low output a high proportion of land was registered [GOSV, 2006]. The link between titling and higher farm output in Vietnam as a result of increased investment was explored in an analysis by Van den Broek et al of data from a sub-survey within the *Access to Resources Household Survey* in 2006. The sample in the sub-survey comprised 2,234 rural households. In their study, they compared the rice production of those possessing land title (over 80 per cent of the sample) with the remainder who did not [Van den Broek et al, 2007]. Their conclusion was that the households with title had significantly higher output per head than those without. The greater increase in crop yield on titled land compared to non-titled land, and in communes where a greater proportion of land was titled, suggests that titling may have led to greater investment in the land through erection of storage buildings, installation of irrigation and drainage systems, the use of labor saving equipment, and the purchase of better seed strains, pesticides and fertilizers, as well as improved land management [Van den Broek et al, 2007]. For example, Van den Broek et al cite figures to show that 74 per cent of registered land holdings had irrigation systems compared to 51 per cent of those which were not titled [Van den Broek et al 2007]. One reason may be that that the conferment of title has led to a greater sense of ownership and security, which in turn provides the motivation and confidence to invest in the land even in the long term.

A recent micro-study of land titling in four communities in Chuquisaca in Bolivia, showed that whilst it had little impact on investment in working capital, it was key factor in increasing investment in perennial tree crops, a key fixed asset. Of particular significance was that amongst those interviewed nearly all who had title had replaced their old tree crops with higher yielding and better quality varieties, whilst most without title retained their old lower quality tree crops even when their yield was declining as a result of age [De Mey and Bertón, 2004].

It has been argued that informal tenure based on and customary rules and community sanction may guarantee equal security of land rights as legal title, and so is no less a spur to long term investment [Stanfield, 1990; Melmed-Sanjak and Lastarria-Cornhiel, 1998]. However, where land has a high commercial potential and is much sought after by big business interests for ranching, logging, plantation farming, mining, and resort development, the protection offered by informal tenure may count for little. In such cases, informal landholders whether occupying land under individual or communal tenure may be dispossessed or find stretches of their land encroached upon. Without legal redress, little can be done to rescind the claims of big business and recover the land lost. This has been all too often the fate of small informal farmers and shifting cultivators in the forested areas of Kalimantan and Sumatra in Indonesia and in Sarawak in East Malaysia, where large stretches of forest land have been appropriated under lease concessions from the state for commercial logging and oil palm cultivation in large plantations. Where land has commercial value threatening the tenure of small holders, legal title is the all the more important in providing the security for long-term investment [Colchester et al, 2007].

Increasing Access to Credit

It has been widely acknowledge that land titling increases the flow of credit to farm enterprises, which further facilitates investment in essential capital inputs. Firstly, it increases the demand for credit. With secure title and little chance of being dispossessed, farmers are more willing to undertake long term investment planning and to take out loans to finance the investment. Secondly, land titling increases the availability of credit. Banks and other lending institutions are more willing to lend to farmers with land title, given that they are unlikely to lose their land and so end up in circumstances where they could default on their repayments. In addition, since the borrower is the legally recognized owner, titled land can be accepted as collateral. In the event of default, the bank or lending institution knows that its claim against the land is legally valid, so allowing it to repossess it and recoup the loss. This thus reduces the risk of lending and so reinforces the willingness to make loans available to farmers. With such guarantees, banks and lending institutions are prepared to allow less onerous terms of repayment, permitting a longer period of amortization, lower rates of interest and smaller deposits. Furthermore, once land is titled, it's market value increases, and it can fetch a higher price than equivalent untitled property. This allows the value of the collateral to increase, with the result that more money can be borrowed to finance farm investment [WB, 2003a; Feder et al, 1988; Carter and Olinto, 2003; Burns, 2006].

For example, It has been widely acknowledged that land titling has resulted in greater credit acquisition by Thai farmers to enable investments in the land to be financed (such as in drainage and irrigation systems). In one evaluation study in 2004, it was stated that apart from increased access to credit, title holders were allowed lower rates of interest and could borrow from a greater range of lending institutions. The value of loans to title holders was between 75 per cent and 123 per cent higher than the value of loans to farmers without title [Burns, 2004]. In the recent micro-study of four rural communities in Bolivia mentioned above, farmers with title were able to access credit whilst most of those without title were denied it. This was in the main due to the insistence of the main rural credit institution Asociación Nacional Ecuménica de Desarrollo that it would advance loans for farming purposes on the basis of title or an equivalent legally recognized proof of ownership [De Mey and Bertón, 2004].

Expansion of a Land Market

A further effect of land titling which has a direct impact on agriculture in the developing world is to encourage the development of a land market. All too often, in small farm communities of developing countries, the selling and buying of farm holdings is the exception rather than the norm. This has hindered the transfer of land to more progressive commercially-minded farmers, committed to adopting modern farm practices in order to raise output and household income.

Land titling is considered an important step in opening up a land market in small farm communities, so that more land can be appropriated by progressive farmers. When a property has title, a purchaser knows that the vendor is the legal and undisputed owner, which means that if a sale is agreed, the transaction is legally valid and he/she will become the legal and undisputed owners. Without that guarantee, the purchaser cannot be fully certain that the vendor is the legal owner and that on acquiring the land his/her legal ownership is beyond dispute. Furthermore, the title registration details provides the purchaser with other important information about the property such as boundaries, previous transfers of ownership, easement and usufruct rights, and rent charges and other burdens attached to the land. Equipped with such guarantees and other information provided by title registration, farmers with ambitions to increase their output and agro-business eager to establish a commercial farm enterprise are more ready to purchase extra land.

In Thailand there is evidence of the emergence of a land market as a consequence of land titling. During the first two phases of the titling program, the price of agricultural land rose nearly threefold, with the price of titled land 75% to 197% higher than the price of non-titled land [Burns, 2004]. With the granting of land titles, land has become a more readily tradable commodity, allowing land to be transferred from farmers who do not wish or who are unable to continue working the land, to those who seek to expand their farm enterprises. In addition, as a consequence of titling, banks in Thailand have become more willing to accept newly purchased land as collateral for a mortgage by which the purchase is financed [Leonard and Ayutthaya, 2003].

FACTORS LIMITING THE BENEFITS OF LAND TITLING

In some cases the beneficial impact of land titling on farm output and incomes has not been fully realized. This is due to the absence of other conditions which are required in conjunction with titling in order to achieve higher output and increased incomes. The absence of these conditions is especially evident in the small farm sector of developing countries and will be discussed below [Holstein, 1996].

Limited Means to Acquire Capital in the Small Farm Sector

One factor limiting the impact of land titling is the size of holdings in the small farm sector, and the consequent restriction on the means to acquire capital. Over much of the developing world average holding size is below 3 hectares, with many landholders occupying less than one hectare. For example, a return provided by the Ministry of Agriculture and Cooperatives in 2006 in Thailand, showed that in 28 out of 75 provinces average holding size was below 20 rai (3.2 hectares). In fact, in five of these provinces the average size was below 10 rai (1.6 hectares) [Office of Agricultural Economics of Thailand, 2006]. In Vietnam, the holding size was even smaller. An official nation-wide survey of agriculture in 2006 shows that the average area of holdings of nearly 10 million agricultural households is 0.63 hectares with more than 60 per cent of households occupying less 0.5 hectares [GSOV, 2006b]. In China, according to surveys from the second national agricultural census of 2006/07, the number of agricultural households is just over 200 million, with an average holding size of 0.61 hectares [National Bureau of Statistics, China (NBSC), 2008a; NBSC, 2008b]. In India, according to the last agricultural census of 2000/01, the average size of 120 million holdings was 1.33 hectares [Department of Agriculture and Cooperation, India, 2001].

The simple fact is that working on such a restricted acreage, farmers in these and many other developing countries find it difficult to eke much more than a basic subsistence living. After land is set aside for producing staple crops for every day food needs, little land is left for producing cash crops for the market. With such a small money income, farmers can hardly afford to purchase expensive but necessary capital inputs on which improved output in the small farm sector depends [Stanfield, 1990; Melmed-Sanjak and Lastarria-Cornhiel, 1998].

Inadequate Extension Services

Land titling also depends upon adequate agricultural extension services in order to stimulate higher investment and output in farming. Through extension outreach, farmers are informed of the existences of new practices and technology in agriculture, and given training and advice on how to adopt these practices and use such technology. In addition, extension officers seek to instill in small farmers a more receptive attitude to new practices and technology, in view of their often strong attachment to traditional methods of cultivation and livestock husbandry and the lack amongst many of the motivation to maximize their return from the land. Due to capacity constraints, resulting from the inadequate numbers of professionally trained agricultural advisers, restrictions on budget resources and lack of properly managed extension agencies, the extension services in many developing countries are not able to meet the need for training and advice especially for small farmers [Rogers, 1996; Van Crowder et al, 1999]. An example is Vietnam where 8,400 extension officers were employed in 2006, with one extension officer on average covering 1,200 farm households. In fact only 26.5 per cent of village level districts (with an average population of just under 470 agricultural households) had an extension officer, and in some provinces there was hardly any village with an extension officer [GSOV, 2006]. Yet despite the limited numbers of extension officers, in 33 per cent of communes, farmers attributed improvements in food crop and perennial crop to the advice and support of the agricultural extension service, with 36 per cent In the case of industrial and perennial tree crop production, this figure increases to 36 per cent [GSOV, 2006].

Inadequate Market Infrastructure and Lack of Farm Insurance

A further requirement for the benefits of land titling to be realized is the existence of an effective market infrastructure. This entails, amongst other things, good roads and mechanized transport linking farms to markets, proximity of market centers, proper facilities for the storage and preservation of produce at market centers, contacts with large volume buyers, and the use of purchase contracts. Also important is market information, by which prices can be set by farmers in light of a full knowledge of market demand. Such requirements are often not met in the small farm sector in developing countries so limiting the opportunities of small farmers to market their produce [Eaton and Shepherd, 2001; Tracey-White, 2005].

However, in recent years, steps have been taken to improve marketing in some developing countries.. One has been the introduction of cooperative arrangements in small farm communities especially in such areas as transportation, the use of market facilities, selling to large volume buyers, and the gathering of market information. Another improvement has been the expansion of contract purchasing, under which farmers enter into an agreement with a large volume buyer such as a food processing company, wholesaler, exporter, supermarket, or hotel to purchase on a regular basis produce meeting a quality standard at an agreed forward price. In return the contract purchaser provides the farmer with technical advice and capital inputs [Eaton and Shepherd, 2001; Tracey-White, 2005; Marocchino, 2009]. An example is the advanced contract purchasing of sugar in Thailand, involving small sugar cane cultivators and milling companies. The government regulates the sharing of the revenue with 70 per cent apportioned to the growers and 30 per cent to the milling companies [Eaton and Shepherd, 2001].

In view of the risky and uncertain nature of farming, it is necessary that farmers have insurance cover. This can protect them against a sharp drop in earnings in the event of animal disease and injury, crop and tree disease, pest infestation of crops, harvest failure due to unusually adverse weather such as drought, and flooding of farm land. A newly titled small farmers with only small cash reserves, investing in their land and obtaining credit, have particular need for insurance to protect against these eventualities which may cancel out any return on the investment and prevent repayment of the debt. But affordable insurance schemes for small farmers in developing countries whether provided by the state or by insurance companies are either not available or not availed of, further undermining the potential benefits of land titling [Eaton and Shepherd, 2001; Motiram and Vakulabharanam, 2007].

Variation in the Impact on Credit Supply

Whilst land titling generally increases farmers' access to credit necessary to investment and land purchase, two problems have arisen: sometimes the credit made available either falls short of the needs of the farmer or is in excess of those needs. These two problems will be discussed below

With respect to the first problem, a number of reports have indicated that small farmers have struggled to obtain even small loans even when their land is registered. This may be due to limited availability of credit and lack of liquidity in the banking system of a country, and the assessment of lending institutions that small-scale loans to farming are not cost effective. Another concern that, in the event of default and the repossession of the land, the outstanding balance of the loan cannot be recouped due to a fall in the market value of the land. Even worse for the creditor is the possibility that eviction and repossession are not a realistic option, especially when the landholder relies on the land for his/her primary subsistence and has no alternative way to earn a living. This makes it difficult for creditors to repossess the land without appearing to the community and the government as inflicting hardship and suffering, and without provoking local hostility [Stanfield, 1990; WB, 2003a; Melmed-Sanjak and Lastarria-Cornhiel, 1998]. Referring to credit access of title farmers in Africa, one FAO report stated:

> Second, with regard to credit access, financial markets in Africa are generally not available for small farmers, even if they hold titled land. This is mainly because transaction costs for small farmers are prohibitively high and because bank credit going to agriculture is determined by factors other than profitability. In other words, the credit for agriculture is not available in any case. In addition, land as collateral is not that attractive or useful in Africa since, because of cultural norms and administrative problems, financial institutions are loath to foreclose on landed property [Melmed-Sanjak and Lastarria-Cornhiel, 1998].

The opposite problem may also arise, viz. that in countries where there is a great deal of liquidity in the banking system, the flow of credit to the farming sector in the wake of a land titling program could become excessive. There is then a risk that small farmers may borrow beyond their means. This may occur if the loans are used to finance capital inputs which are not cost effective in light of the nature and size of the farm enterprise, or are used to pay for consumption of goods and services, such as consumer durables and medical and healthcare outlays. The risk is all the greater when small farmers are subject to sharp drops in income as a result of variable harvests and fluctuating prices. The result is that in order to repay their loans, small farmers sacrifice future capital investment or forgo necessary consumption. In the event of default, the farmer may lose some or all of his/her land accepted as collateral. To avoid this outcome, farmers may turn to informal moneylenders to pay the outstanding loan, so creating additional debt often at usurious rates of interest. This may in the long term exacerbate the degree of indebtedness.

An example of a land titling program that gave rise to excessive borrowing by certain small farmers is that in Thailand, despite the overall benefits of that program in fostering the development of agriculture. One report indicated that during the 1990's which was the middle phase of Thailand's land titling program, debt amongst new title holders shot up by 40-60 per cent. A survey of 50 villages in Northern Thailand showed that on average a peasant household was in debt to the tune of 70,000 Thai Baht (US$1,600). This was no small amount amongst poor semi-subsistence rural dwellers given an estimated average yearly income of 12,000 Thai Baht (US$288), an income to loan multiple of 5.5 [Leonard and Ayutthya, 2003]. A more comprehensive enumeration in 2005 found that average debt outside Bangkok by region, including urban dwellers, ranged from 58,000 Thai Baht (US$1,325) to 76,000 Thai Baht (US$1,737) [Agricultural Land Reform Office (ALRO), 2006]. Often the loans were used to finance consumption, children's education or healthcare, or to repay other debts. In other cases, the money was borrowed to fund investments that turned out to be non-productive or loss making. Such loans could only be repaid and serviced through existing (often small) income stream to the family whether from selling farm produce, part-time or casual employment, or remittances from family members working elsewhere,.

With a high loan to income ratio for low income households, not surprisingly, many farmers in Thailand struggled to meet their debt repayments. The consequence was to squeeze existing income, which exacerbated the poverty of the household and prevented the acquisition of necessary capital inputs such as fertilizers and seeds to sustain the productive output of the farm [Leonard and Ayutthya, 2003; Slaats, 1999]. Where the household fell behind in repayments and were in danger of default, the practice was to turn to usurious money lenders, who charged much rates of interest to acquire the finance to repay the institutional loan. This has further added to the hardship of indebted households. In the worst cases, they have been let little option but to sell their land and thus lose their chief means of subsistence [Leonard and Ayutthya, 2003; Slaats, 1999]. The impact of indebtedness on poverty can be gauged from the feedback given by eight million poor people in the consultation exercise conducted by the National Poverty Eradication Centre in 2005. Five million of them reported that their biggest single problem was debt [ALRO, 2006].

To avoid a credit bubble as a consequence of land titling, it is necessary for lending institutions to tailor advances to small farmers on the basis of careful assessments of the benefits of the investment to be financed, the means available for repayment, the risks of default arising from the variability of yield and prices, with additional protection from sizeable deposits required before the loan is advanced. In Vietnam, lending institutions have adopted restrictive conditions on lending. As a result a credit bubble did not follow land titling as it did in Thailand. According *Vietnam Household Living Standards Survey*, 45 per cent of farming households were in debt in 2006, with the amount of money borrowed per household little more than the average yearly income [GSOV, 2006; Iyer and Do, 2008].

Impact of Land Titling on Agriculture in Tribal Communities

The question arises whether land titling has a beneficial impact on the development of agriculture when applied to traditional tribal communities occupying land under communal tenure, and practicing sedentary farming combined with common grazing, nomadic herding, or shifting cultivation, referred to above. There are different perspectives on this issue.

When collective title is awarded to tribal communities, this will underpin their traditional communal way of life and will enable them to continue farming practices which they have followed over many generations and of which they have intimate knowledge. Of particular importance is the protection afforded against land encroachment by big business for purposes of mining, resort development, logging, and plantation farming. The vulnerability of tribal and ethnic minorities who lack land title, has been illustrated in a number of countries in recent years. In Sarawak in Malaysia, Dayek tribal peoples have suffered the loss of many hectares of forest land to logging companies awarded concessions by the Sarawak sate government. Both the Sarawak state government and the federal government in Malaysia have refused to recognize their title even though their title claims have been upheld by the Malaysian federal and state courts on the basis of common law principles [Colchester et al, 2007]. In Nepal the Chepang , Sherpa, Rai and Shingsa indigenous nationalities in central and eastern Nepal, who practice shifting cultivation on steep upland slopes, have suffered from the refusal of the government to title their land, and have in recent years lost large areas for forest and other types of commercial plantation [Aryal and Kerkhoff, 2008]. Likewise, in Peru, Amazonian tribal groups, and in Botswana, the Bushmen of the Kalahari, who both lack full title to their communal lands, are currently faced with the prospect of losing areas they occupy to mining companies. In many of these cases, protests and violence have occurred [Kumar, 2009; Navai, 2009; Survival International, 2009].

A contrasting perspective is that collective titling of communal tribal land without creating strong individual land rights, is detrimental to both land conservation and agricultural output. The detrimental effects are associated with three main forms of land use in tribal farming, viz. common grazing, nomadic herding, and shifting cultivation.

In the case of tribal land used for common grazing and nomadic herding, collective titling may encourage the continuation of these practices, the consequences of which are often over-grazing, spread of livestock disease and a lack of control over breeding. The result is land degradation and a decline in the quality of livestock. The argument against the collective titling of tribal land used for shifting cultivation, likewise, rests upon the detrimental consequences that may arise by permitting such practices to continue. Shifting cultivation is often considered as not conducive to long term investment and high output, so confining its practitioners to a largely subsistence livelihood. It is reasoned that cultivators moving every so many years from one piece of land to another have no inducement to invest in and improve the land as they would if they occupied a permanent holding.

It is also argued that collective titling of tribal land, whether used for grazing or shifting cultivation, with a restriction on selling land to non-members of the tribe, would impede the transfer of land to progressive farmers most able to develop it for farming purposes. What's more, collective titling may prevent areas of land, which have commercial potential for mining and plantations, from being exploited so that they remain an under-utilized and wasted resource.

In order to avoid these outcomes and foster the commercial development of tribal lands, governments have often either refused to recognize collective tribal title, or have granted title on a modified or individual basis. For example, in Vietnam as mentioned above, tribal households who had engaged in shifting cultivation, have been allotted permanent holdings on the tribal lands, together with a certificate of title, under the condition that they invest in long-term tree crops yielding a commercial return [Novellino, 2001]. A similar arrangement has been implemented in Laos [Soulivanh et al, 2004]. Likewise, in Botswana as mentioned above, certain areas of common tribal grazing were divided into large ranches in the 1970's, under a lease title, with the aim of preventing overgrazing, improving breeding and livestock management, and establishing a commercial and export-oriented beef industry. The greater commercial orientation of the ranch farm sector in Botswana is reflected in extent to which the cattle off-take was higher than in the communal grazing sector. Only in 3 years between 1979 and 2004 was the rate of off-take in the ranch sector lower than that in the communal grazing sector, and in most years was twice or three times higher (in one year it was nearly four times higher) [GOB, 2007].

However, whilst the granting of individual title on tribal lands may have spurred commercial progress, challenges have arisen. In Laos and Vietnam, with the consequent restrictions on communal tenure and shifting cultivation, the concern has focused on the erosion of communally based way of life and system of farming within the tribal communities, and the reduced access to land for tribal households [Novellino 2001]. In Botswana, in certain areas ranchers have continued to over-graze the pasture and have encroached upon the communal pasture reserves used by small cattle owners [Darkoh and Mbaiwa, 2002; Knox, 1998b, 215; Frimpong, 1995].

CONCLUSION

In many developing countries, in recent years, governments have promoted systematic land titling programs, entailing the legal recognition and official recording of the ownership of parcels of land with defined boundaries. The reason for such programs is the widely acknowledged benefits of titling in boosting agricultural investment and output and thereby raising farm incomes. Of particular importance are the incentives given to long-term investment arising from greater sense of ownership and security, the increased access to credit with land used as collateral, and the increased marketability of land allowing land to be transferred to progressive farmers. The titling programs have been facilitated by technological advances in surveying and mapping, the adoption of on-line systems of title registration, and the use of community level adjudication based on interviews of title claimants, neighbors and community leaders, especially important in the absence of documents and records to prove ownership.

A particular focus of titling programs have been small farmers who often lack title to their land and claim ownership on an informal and customary basis. The lack of legal tenure to their tenure has been seen as providing a disincentive to invest, impeding their access to credit, and hindering the transfer of land to those wishing to expand their enterprises. There is a significant body of evidence to show that small farmers with commercial skills who have received title, have been encouraged to invest and adopt modern methods of agriculture, gaining access to credit and acquiring more land through a more mobile land market.

However, the potential benefits of land titling have in some cases not been fully realized. The key point here is that maximizing the impact of land titling for small farmers entails other requirements. These include the availability of affordable capital inputs, provision of extension services to advise on the adoption of improved methods of farming, the development of appropriate marketing outlets, accessibility to micro-credit schemes, control over the supply of credit, and adequate insurance coverage to protect against the risks and uncertainties of farming. All too often such conditions essential to upgrading small farms have not been met during the implementation of systematic titling programs. The lesson to be drawn is that land titling should be one element in a wider and coordinated program providing a wide range of services to enhance the commercial status of newly titled small farms. This necessitates coordination, which is often lacking, between the land registration authority and other central and local government agencies with responsibility for agricultural development.

However, dilemmas arise in the titling of land occupied by tribal or indigenous communities. Collective titling, if adopted, helps to preserve the traditional way of life and working practices of the tribe. But it may restrict the scope for commercial exploitation of land which has productive potential., and, by not according individual title, may reduce the possibility of tribal households becoming commercial farmers. The alternative is to grant individual title to tribal households or to grant individual title as a derivative of collective title, with in both cases, areas of land reserved for communal use and commercial exploitation. If either is chosen, commercial development of tribal land still depends, as with titling of non-tribal land, on the provision of necessary state support and services in relation to capital, extension, marketing and insurance.

REFERENCES

Agricultural Land Reform Office (2006). *National report on agrarian reform and rural development in Thailand.* Bangkok: Ministry of Agriculture and Cooperatives.

Antle, J., Yanggen, D., Valdivia, R., and Crissmand, C. (2003). *Endogeneity of land titling and farm investments: Evidence from the Peruvian Andes.* Bozeman Montana: Department of Agricultural Economics and Economics, Montana State University.

Aryal, K.P., and Kerkhoff, E.E. (2008). *The right to practice shifting cultivation as a traditional occupation in Nepal.* Geneva: International Labour Organization.

Besley, T. (1995). Property rights and investments incentives: Theory and evidence from Ghana. *Journal of Political Economy*, 103(5): 903-37.

Binns, B., and Dale, P. (1995). *Cadastral surveys and records of rights in land.* FAO *Land Tenure Studies 1*. Rome: FAO.

Burns, T. (2004). *Thailand's 20 year program to title rural land, Background paper prepared for the World Development Report 2005*. Washington: World Bank.

Burns, T. (2006). *Land administration reform: Indicators of Success and. future challenges. Agriculture and Rural Development Discussion Paper 37*. 2nd edition. Washington DC: World Bank.

Carter, M., and Olinto, P. (2003). Getting institutions right for whom? Credit constraints and the impact of property rights on the quantity and composition of investment. *American Journal of Agricultural Economics*, *85*: 173-86.

Cleary, M., and Eaton, P. (1996). *Tradition and reform: land tenure and rural development in Southeast Asia*. Oxford: Oxford University Press.

Chemonics International Inc (2007). *Bolivia Land Titling Program Quarterly Report: July September 2007*. Washington: United States Agency for International Development.

Congress of the Philippines (1997). *Indigenous Peoples Rights Act*, Republic Act No. 8371.Manila: Author.

Darkoh, M., and Mbaiwa, J. (2002). Globalisation and the livestock industry in Botswana. *Singapore Journal of Tropical Geography, 23*(2): 149-166.

De Mey, K., and Bertón, B. (2005). *Influence of land titles on agricultural production and rural livelihood development in the Canyon of Camargo, Chuquisaca – Bolivia.* Copeen: Department of Economic and Natural Resources The Royal Veterinary and Agricultural University.

Department of Agriculture and Cooperation, India (2001). *Agricultural census, 2000/01.* Delhi: National Informatics Centre.

Eaton, C., and Shepherd, A.W. (2001). *Contract farming: Partnerships for growth, FAO Agricultural Services Bulletin 145.* Rome: FAO.

Evrard, O. (2004). *La mise en oeuvre de la réforme foncière au Laos: impacts sociaux et effets sur les conditions de vie en milieu rural.* LSP Working Paper 8. Rome: FAO.

Ezigbalike, C., Rakai, M.T., and Williamson, I.P. (1996). *Information systems cultural issues in land: Pt 1, Land Reform Bulletin 1996.* Rome: FAO.

Feder, G., Onchan, T., Chalawong, Y., and Hongladarom, C. (1988). *Land policies and farm productivity in Thailand.* Baltimore, MD: Johns Hopkins University Press.

Frimpong, K. (1995). A review of the tribal grazing land policy in Botswana. *Pula: Botswana Journal of African Studies. 9* (1): 1-15.

Fujita, Y., and Phanvilay, K. (2004). Land and forest allocation and its implication on forest management and household livelihoods: Comparison of case studies from CBNRM research in Central Laos. Paper presented to the Tenth Biennial Conference of the *Association for the Study of Common Property. Oaxaca,* México, 9 – 13 August.

General Statistics Office of Vietnam (2006). *Results of the 2006 rural, agricultural and fishery census: volume 1 - general results.* Hanoi: Author.

Government of Botswana (2007). *2004 Botswana agricultural census report.* Gaberone: Central Statistics Office.

Government of the Socialist Republic of Vietnam (1993). *Law on Land, 1993.* Hanoi: National Assembly.

Government of the Socialist Republic of Vietnam (2003). *Law on Land, 2003: No. 13-2003-QH11.* Hanoi: National Assembly.

Ha, T.V. (2007). *Impacts of land allocation and land titling on rural livelihoods: An examination from the southern uplands of Vietnam.* Paper given to International Agricultural Trade Research Consortium, Symposium Beijing, July 8-9.

Heryani, E., and Grant, C. (2004). Land administration in Indonesia. Paper given to 3rd FIG Regional Conference, Jakarta, Indonesia, 3-7 October.

Holstein, L. (1996). *Towards best practice from world bank experience in land titling and registration,* Washington: World Bank.

Iyer, L., and Do., Q (2008). Land titling and rural transition in Vietnam, *Economic Development and Cultural Change. 56* (3): 531-580.

Kumar, C. (2009). Peru's political fault line. *Guardian* (London), 22 June.

Knox, A. (1998a). Ghana country profile. In John Bruce (Ed.). *Country profiles of land tenure: Africa, 1996.* Madison: Land Tenure Centre, University of Wisconsin-Madison: 68-73.

Knox, A. (1998b). Botswana country profile. In John Bruce (Ed.). *Country profiles of land tenure: Africa, 1996.* Madison: Land Tenure Centre, University of Wisconsin-Madison: 212-217.

Leonard, R., and Ayutthya, K. (2003). *Thailand's land titling programme: Securing land for the poor*. Oakland CA: Land Action Research.

Marocchino, C. (2009). *A guide to upgrading rural agricultural retail markets. Agricultural Management, Marketing and Finance Working Document 24*. Rome: FAO.

McEwen, A. (2001). *The significance of land title registration: a global perspective*. Paper presented to the 75th Annual Meeting of the Institute of Surveyors of Sri Lanka, Colombo, September.

Melmed-Sanjak, J., and Lastarria-Cornhiel, S. (1998). *Land access, off-farm income and capital access in relation to the reduction of rural poverty. Land Reform Bulletin 1998/1*. Rome: FAO: 1-15.

Motiram, S., and Vakulabharanam, V. (2007). Corporate and cooperative solutions for the agrarian crisis in developing countries. *Review of Radical Political Economics, 39*: 360 – 367.

National Bureau of Statistics, China (2008)a. *Communiqué on major data of the second national agricultural census of China (no.1) 2008*. Beijing: Author.

National Bureau of Statistics (2008)b. *Communiqué on major data of the second national agricultural census of China (no.6) 2008*. Beijing: Author.

Navai, R. (2009). Peru Indian tribes join forces to fight off Amazon sale to oil companies. *The Times* (London), 9 October.

Ngo, V. L. (1993). Reform and rural development: Impact on class, sectoral, and regional inequalities. In William Turley and Mark Selden (Eds.). *Reinventing Vietnamese Socialism*. Boulder, C.O.: Westview Press: 165-208.

Novellino, D. (2001). *Indigenous highlands in transition: The case of Ha Giang Province in Northern Viet Nam. Land Reform Bulletin 2000/2*. Rome: FAO.

Office of Agricultural Economics (2006). *Land use of agricultural sector of Thailand in national year 2549*. Bangkok: Ministry of Agriculture and Cooperatives: available at http://www.oae.go.th

Ravallion, M., and Van de Walle, D. (2003). *Land allocation in Vietnam's agrarian transition, part 1: Breaking up the collective farms. IFS Working Paper, EWP03/03*. London: Institute for Fiscal Studies.

Rogers, A. (1996). *Participatory training: Using critical reflection on experience in agricultural extension training. Economic and Social Development Series (54)*. Rome: FAO.

Slaats, H. (1999). Land titling and customary rights: Comparing land registration projects in Thailand and Indonesia. In Toon van Meijl, and Franz von Benda-Beckmann (Eds.). *Property rights and economic development : Land and natural resources in Southeast Asia and Oceania*, London: Kegan Paul: 88-109.

Soulivanh, B., Chantalasy, A., Suphida, P., and Lintzmeyer, F. (2004). *Study on land allocation to individual households in rural Areas of Lao PDR*. Vientiane, German Technical Cooperation.

Stanfield, D. (1990). *Rural land titling and registration in Latin America and The Caribbean: Implications for rural development programs*. Madison, Wisconsin: Land Tenure Center, University of Wisconsin.

Survival International (2009). *The Bushmen*. London: Author: available at http://www.survivalinternational.org.

Suttie, J., Reynolds, S., and Batello, C. (2005). *Grasslands of the world.* Rome: FAO.

Tracey-White, J. (2005). *Rural–urban marketing linkages: An infrastructure identification and survey guide. FAO Agricultural Services Bulletin 161.* Rome: FAO.

Van Crowder, L., Lindley, W.I., Bruening, T., and Doron, N. (1999). *Agricultural education for sustainable rural development: challenges for developing countries in the 21st century.* Rome: FAO.

Van den Broeck, K., Newman, C., and Tarp, F. (2007). *Land titles and rice production in Vietnam.* Dublin: Department of Economics, Trinity College Dublin.

World Bank (2003a). *Land policies for growth and poverty reduction: A World Bank policy research report.* Oxford: Oxford University Press.

World Bank (2003b). *Comparative study of land administration Systems – Thailand case study.* Washington DC: Author.

Yokoyama, S., Tanaka, K., and Khame, P. (2006). *Forestry policy and swidden agriculture in Laos.* Paper given at SEAGA Conference, Singapore, 28-30 November.

Zevenbergen, J. (1998). *Is title registration really the panacea for defective land administration in developing countries?* Paper given to International Conference on Land Tenure in the Developing World, Cape Town, 27-29 January.

In: Agricultural Production
Editor: Felix C. Wager, pp. 139-161
ISBN: 978-1-61668-695-6

Chapter 7

SUSTAINABLE IRRIGATED AGRICULTURAL PRODUCTION OF COUNTRIES IN ECONOMIC TRANSITION: CHALLENGES AND OPPORTUNITIES (A CASE STUDY OF UZBEKISTAN, CENTRAL ASIA)

***Shavkat Rakhmatullaev*[*1,2], *Frédéric Huneau*[‡2], *Philippe Le Coustumer*[†2], *and Mikael Motelica-Heino*[±3]**
[1]Tashkent Institute of Irrigation and Melioration (TIIM),
Tashkent, Uzbekistan
[2]Université de Bordeaux, Talence, France
[3]Université d'Orléans, CNRS/INSU, Université François Rabelais - Tours Institut des Sciences de la Terre d'Orléans, France

ABSTRACT

For the fulfillment of the thirsty ambition of self-sufficiency of the Soviets for cotton production, the arid Central Asian region, particularly, Uzbekistan has been extensively exploited. In fact, vast tracts of deserts have been converted into irrigated agricultural lands without proper consideration of the environment and technical standards. As a result, trends in natural resource degradation (soil salinity, desertification, water quality) as well as declining crop yields have dramatically increased.

The agricultural sector is the backbone for employment, food security and export revenues of the Central Asian countries. Since the independence of the Central Asian countries (after the breakup of the former Soviet Union), the situation has changed dramatically in terms of institutional, political and technical systems. Political transition, which is defined as a shift from a once planned centralized economy to a market-driven one, has introduced 'new' concepts like land tenure, water rights and different kinds of ownership. All such transformations have impacted the agricultural production in Central Asia.

The institutional change can be described as decentralization of the farming systems, i.e. transition from the former state collective farms into the smaller forms of private farms. The institutional interventions are aimed to increase agricultural production through improving water management. It is arguable that private production systems are the most effective business-driven forces but the situation is quite different in Central Asia due to the irrigated agriculture.

The biggest challenge for a sustainable irrigated agricultural production lies in the recent reforms of the water management sector in Central Asia and Uzbekistan. Water users' associations have been established to replace the former collective farming systems for irrigation water distribution and maintenance of irrigation infrastructures at the on-farm level. The intention of the national government was to shift the operation, maintenance and management of irrigation infrastructures to non governmental institutions (decentralization). However, these institutions have not fulfilled their promising tasks because of i) a rapid increase in the number of private farms along canals; ii) the cropping structure which is a mosaic with different crop water requirements compared to the former monoculture; iii) poor financial, training and technical capacities of the newly established institutions; iv) a state-ordered agricultural production quota system (for cotton and wheat).

This paper analyzes the historical aspects of the transformation in the farming production institutions in Central Asia with a special focus on Uzbekistan and comprehensively overviews the main current challenges facing the farming system and potential opportunities for reversing the situation.

Keywords: irrigated agriculture; water users' associations; agricultural water management; soil and water pollution; cotton production; Central Asia

1. INTRODUCTION

After the fundamental and extremely fast political transformation of the Central Asian (CA) region (collapse of the Former Soviet Union and emergence of National States), the land-locked arid countries of Central Asia (Kazakhstan, Tajikistan, Kyrgyzstan, Uzbekistan and Turkmenistan) have directed their natural resources utilization for the economic development from a previously central-planned economic system to a market-oriented one. The once interconnected branches of economies in Central Asia have become fragmented and collapsed (UNEP, 2005). Many thousands of jobs have been destroyed. Business and industry sectors with rich oil and gas reserves have shown their flexibility under a new transformation process but the most impacted was agriculture.

In this hard period, agriculture has played a vital role for absorbing employment of rural populations. Kandiyoti (1999) describes the agricultural sector as being a "shock absorber" for the surplus of rural labor. In fact, in the CA region, agriculture employs about 45% of the total population and its share in GDP (Gross National Product) is 24% on average (FAO, 2009). Some 22 million people depend directly or indirectly on irrigated agriculture in these countries (World Bank, 2003). Besides, the export of cotton contributes to significant foreign cash revenues for Tajikistan, Turkmenistan and Uzbekistan. Sustainability of irrigated agriculture is one of the main platforms for food security, employment, livelihoods and environmental protection.

With the independence of the CA countries, the situation has changed dramatically in terms of institutional, political and technical systems (IAMO, 2008). Political transition from a once planned centralized economy to a market-driven one has introduced 'new' concepts like land tenure, water rights and different kinds of ownership. Institutional reforms have been implemented for arrangements of previously state-owned collective farms into a new private form of farming system. Ideally the de-centralization of farming systems should have created new opportunities for rural people. Indeed the fast privatization of agricultural lands has created thousands of workplaces with the complete privatization of agricultural parcels in Kyrgyzstan and long-term leases in the rest of the Central Asian states (Sehring, 2008).

The reform of agriculture was inevitable in the process of political and economic transformation. The private form of farming system would be better and contribute to agricultural production and, as a result, to tax returns for governments. The climatic conditions and limited water resources of the region impose the practice of irrigated agriculture only, which makes irrigation water supply and management the major factors limiting crop yields in the region (Ibragimov et al., 2007).

About 7.9 million ha of the total irrigated lands in Central Asia, of which about half (4.2 million ha) is irrigated, is in Uzbekistan (World Bank, 2005). Over a period of 90 years (1913–2003) the areas under irrigation have increased by 3 times on average in the CA region, e.g., from 4.51 million ha in 1960 to 6.92 million ha in 1980 and to 7.85 million ha in 2000 (UNEP, 2005; Sattarov et al., 2006). This increase was due to a gigantic Soviet hydraulic program through the construction of dams, irrigation canals, pumping stations and various hydraulic facilities. This hydraulic infrastructure is considered as one of the most complex in the world (O'Hara, 2000; Ximing et al., 2000; UNDP, 2007).

Additionally there has been a general warming up in the CA countries on the order of 1–2^0C since the beginning of the 20th century that might have a strong impact on the regional temperatures and precipitation regime but also on natural ecosystems and agricultural crops (Lioubimtseva et al., 2005).

From a technical point of view the major bulk of hydraulic infrastructure (irrigation and drainage) are in poor conditions and government financial programs for the maintenance have been substantially reduced (UNESCO, 2000; Sokolov, 2006; UNDP, 2007).

The aim of this paper is to comprehensively review the two sets of problems that most impact the sustainability of irrigated agricultural production systems in Central Asia with special focus on Uzbekistan. The first set of problems is reforms in agricultural production systems (de-collectivization), i.e. transformation of a former collective farming into current private farming systems. The second set of problems is the on-farm irrigation water distribution and infrastructure management by recently created local institutions such as the Water Users Associations (WUA).

Along with de-centralization in agriculture, water management system was also transformed in Uzbekistan at two levels (Abdullaev et al., 2008). In this paper we will deal with only irrigation water management at the on-farm level. The irrigation water management was transformed from once a territorial-based management to a hydrographic principle at the basin level. The second level is the creation of WUA at the on-farm level. A WUA is a self-managing group of farmers working together to operate and maintain their irrigation and drainage network (only inter-farm or on-farm level) to ensure a fair and equitable water distribution and increase crop yields (Abdullaev et al., 2009; Kazbekov et al., 2009).

At the watershed level main the irrigation water distribution management and hydraulic infrastructure were left for the government for operation and maintenance. At the on-farm level (secondary and tertiary) irrigation canals were shifted to WUAs. However, these institutions have not fulfilled their promising tasks because of i) a rapid increase of the number of private farms instead of a few former collective ones along a single irrigation canal; ii) along a irrigation canal the cropping structure is a mosaic with different crop water requirements, irrigation regimes in comparison to the former monocultural cropping structure;

iii) a poor financial, capacity building and technical capabilities of the new established institutions; iv) a state ordered agricultural production quota system (for cotton and wheat).

Last but not least there are environmental concerns such as land degradation (elevated land salinity and waterlogging) of irrigated lands. Poor incentives of farmers to preserve the irrigation water, physical deterioration of the irrigation and drainage networks exacerbate the sustainability of irrigated agriculture. As a result the crop yields have declined by 50 percent (UNESCO, 2000). The continued soil degradation would jeopardize the sustainability of irrigated agriculture and impact the livelihoods of million people.

2. Uzbekistan

Uzbekistan is located in the heart of Central Asia with a population over 27 million and borders with Kazakhstan in the north, Kyrgyzstan in the north-east, Tajikistan in the east, Afghanistan in the south and Turkmenistan in the south west (Figure1). Administratively, Uzbekistan is divided into 12 provinces (viloyat) and one autonomous Republic of Karakalpakstan. The deserts (Kara Kum and Kyzyl Kum) and plains (Turan and Ustyurt) stretch from south-east to north-west and constitute roughly 80% of the landscape, the mountain systems of western Tian Shan and Pamir Alay occupying the rest of its territory (Shultz, 1949; Irrigation of Uzbekistan, 1979).

The average monthly air temperature for January ranges from +3^{0}C in the south (Termez) to -8^{0}C in the north (Usyurt plateau) and in the summer temperatures can reach 45-49^{0}C. The average annual precipitation in the desert zone is less than 200 mm while in the piedmont and highland zones it can be of 400-800 mm with a maximum of 2000 mm in the mountain areas (Goskompriroda of Uzbekistan, 2005; UNDP, 2007). The annual average evaporation is about 1600-2200 mm. Thus agricultural production is impossible without irrigation.

Figure 1. Location of Uzbekistan and Central Asia

2.1. Water Resources

With an annual rainfall of 100–300 mm and a mean evaporation of 1600-2200 mm, Uzbekistan has a continental climate of a dry mid-latitude desert, characterized by hot summers and cold winters (Shultz, 1949; FAO, 2009). Its climate is largely arid and its water resources are unevenly distributed both in space and time. There is a strong dependency on winter and spring rains and snowmelt from the Tian Shan and Pamir mountains. In fact these mountains are major contributors to the watersheds of the CA countries (World Bank, 2005; Hagg et al., 2007).

The streamflow is characterized by an extreme intra-annual variability and also unevenly spatially distributed (Rakhmatullaev et al., 2009). Two main river basins are found in Uzbekistan, the Amu Darya and the Syr Darya (Dukhovny, 2007). These transboundary rivers satisfy 82% of the total water demand for irrigation, whereas only 18% of the same demand is satisfied by the internal Kashka Darya, Zerafshan and Surkhan Darya rivers in Uzbekistan (Figure 1) (Shultz, 1949; Nezlin et al., 2004). Uzbekistan heavily depends on transboundary sources of water and on average only 9.52km^3 of water resources are generated within the country whereas 94.8km^3 comes from outside (Statistics Committee of Uzbekistan, 2006; Rakhmatullaev et al., 2009).

2.2. Soils

The soils of Uzbekistan vary according to the latitude and altitude zones which in turn are associated with particular climatic conditions and vegetation cover (Figure2 and Table1).

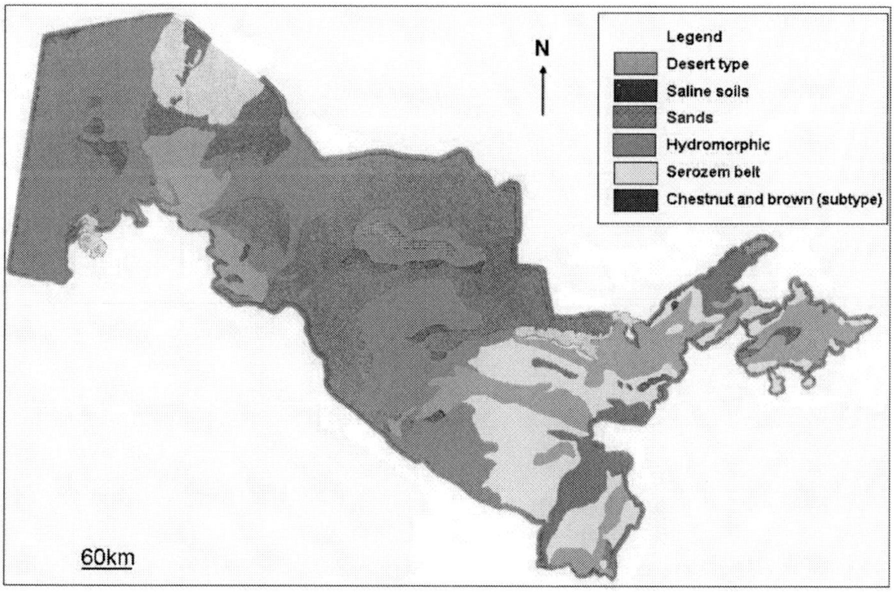

Figure 2. Soil map of Uzbekistan

Due to climatic conditions and aridity about 14.6 million ha (32% of all soil types) are represented by desert type of soils (grey-brownish desert, sandy and takyr with their subtypes). These soils cannot be used for irrigation because of their inherited characteristics, available water resources for irrigation, salinization and harsh relief; they are instead mainly used for cattle breeding of camels and sheep (Talipov, 1992).

Saline soils (solonchak and their subtypes) are distributed in Karakalpakstan and near the Aral Sea area and occupy about 1.3 million of ha (3%) and prevail in the local depressions located in lowland plains, lake basins and between mountains (UNDP, 2007). With appropriate irrigation and drainage systems, they can be used for agriculture. Sands cover more than 12 million of ha (about 28% of the total surface) in Uzbekistan and are unsuitable for agriculture. About a total 63% of soils (out of total of 44.89 million of ha) in Uzbekistan cannot be used for irrigation due to their inherited soil characteristics, available irrigation and drainage facilities.

Table 1. Soil type distribution in Uzbekistan (Talipov, 1992)

Soil type	Area Million ha	% of area occupied to the total	Altitude (above sea level) m
	Plain zone		
Grey-brown desert	11.5	25	150-250
Sandy	1.37	3	120-150
Takyr	1.78	4	120-180
Meadow-takyr	0.47	1	120-150
Meadow and wetland-meadow	1.85	4	80-100
Solonchak	1.27	3	80-100
Sands	12.1	28	120-150
	Piedmont and mountain zone		
Light serozem	2.59	6	250-500
Typical zerozem	3.05	7	500-700
Dark serozem	1.06	2	750-1200
Brown and brownish middle-altitude	1.66	4	1200-2800
Light brownish high-altitude	0.54	1	2800-3500
Meadow-serozem	0.78	2	250-500
Meadow and wetland-meadow	0.75	2	250-500
Rock	3.0	6	-
Water surface	1.12	2	-
Total	44.89	100	

The hydromorphic soils (meadow, meadow-takyr and their subtypes) are distributed over 3.8 million ha (9% of the total surface) in all regions of Uzbekistan but can be found particularly in the middle and lower river reaches of Amu Darya, Syr Darya, Zerafshan, Kashka Darya, Surkhan Darya, and in the Aral Sea wetlands. These soils have a high humus content of 2.5-10% with high absorption capacity and used widely for rice cultivation (Goskompriroda of Uzbekistan, 2005).

The soils of the serozem belt (light, typical and dark) are distributed over 6.7 million ha (15%) on the lower margin of the piedmont plains with an altitude of 200 to 700-900 m above sea level. These soil types have a high humus content (2-4%) and are less subject to salinization and can be used for rain-fed and irrigated agriculture. With the application of agronomic measures (e.g. fertilizers and conservation tillage) these soils can be used more productively for agriculture (Statistics Committee of Uzbekistan, 2006).

About 2% of the land surface in Uzbekistan is occupied with water bodies (reservoirs, lakes, rivers) and 6% by rocks. About 5% of the land (2.2 million ha) are represented by chestnut, brown and light brown soil types in mountains in altitude from 1200 to 1600 m above the sea level and subjected to erosion. The main usages of these soils are plantation of trees and pastures (UNEP, 2005).

2.3. Irrigated Agriculture and Cotton Production

The total cultivated land in Uzbekistan is estimated to be 5.2 million of ha, of which 4.2 million ha are irrigated (UNESCO, 2000). Figure3 depicts the evolution of the irrigated lands in Uzbekistan for the last century. The irrigated lands almost increased by 3 fold over the last century: about 1.3 million ha in 1900, 2.6 million ha in 1950 and 4.2 million in 2000 (UNEP, 2005). However, Uzbekistan has witnessed a population boom; in fact there was an increase of 4 fold of the total population i.e. from 6.5 million in 1940 to over 26 million in 2007. As a result, the irrigated lands per capita reduced from 0.41 ha/person in 1940 to only 0.16 ha/person in 2008 (UNDP, 2009) (Table2).

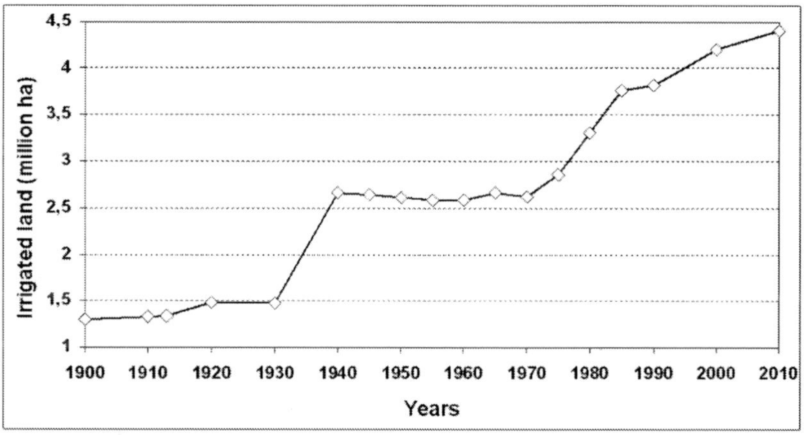

Figure 3. Evolution of irrigated lands in Uzbekistan for the last century

Table 2. Population and per capita irrigated lands transformation in Uzbekistan for the last century (Talipov, 1992; UNDP, 2009)

Year	Total Population ('000 persons)	Per capita irrigated land (ha/person)
1940	6551	0.41
1960	8722	0.30
1970	11799	0.22
1980	16158	0.20
1990	19906	0.19
2000	24487	0.17
2007	26663	0.16

This expansion is supported by the complex irrigation hydraulic infrastructure consisting of more than 14,000 different scale pumping stations, over 28,300 km of inter-farm ditches and main irrigation canals and more than 137,000 km of on-farm irrigation ditches, 42 water diversion structures (with a capacity of 10-300 m^3/sec), 55 water reservoirs and about 30,000 km of drainage systems (Royal Haskoning, 2001) (Table3).

Table 3. Irrigation network of Uzbekistan (Royal Haskoning, 2001; UNDP, 2007)

Hydraulic structures	Location in the system	Unit
Irrigation canal	Main	7923 km
	Inter-farm	20 437 km
	On-farm (include)	137 385 km
	Concrete lined	12 103 km
	Chute	21 668 km
	Conduit	3308 km
Pumping station	Inter-farm	1466 pieces
	On-farm	12 780 pieces
Drainage system	Inter-farm	29 939 km
	On-farm (include)	
	Open	106 321 km
	Closed	30 242 km
	Vertical drainage wells	4309 pieces
Irrigation well		5022 pieces
Water reservoir		55 pieces
Water intake		42 pieces

In order to increase the area under agricultural production, large scale hydraulic projects were started in soviet Uzbekistan. For example, in 1939 during 17 days, an irrigation canal named Lyagan with a total length of 32 km was manually dug with the daily participation of more than 13 000 – 14 000 persons (Talipov, 1992). The Big Ferghana irrigation canal which stretches over 270 km was dug in less than 45 days with the assistance of 160 000 persons (Rakhmatullaev et al., 2003).

In later periods, there have been gigantic projects to construct irrigation canals, water reservoirs, pumping stations and other hydraulic structures with a strong support from Moscow by enormous financial and technical resources. However, environmental concerns were ignored by the centralized Moscow administration and today several known ecological catastrophes such as the Aral Sea catastrophe are witnessed. The detailed discussion on the historical evolution of irrigation in Central Asia can be found in O'Hara (2000).

Talipov (1992) reports that the Soviet Union imported about 80% of its cotton products in 1920s and in order to be self-sufficient for its textile industry by internal cotton production, the Central Asian region was used as a cotton producing platform. The area sown under the cotton increased by 7.5 fold from 1924 to 1990. Figure4 depicts the evolution of area sown under cotton and its yield in Uzbekistan for the last century. For example, in 1913 the area sown under cotton constituted roughly 35% (out of total sown area) whereas at the end of 1990, it accounted to more than 56%. The area under cereals was reduced from 48% (out of total sown area) in 1913 to only 13% in 1990.

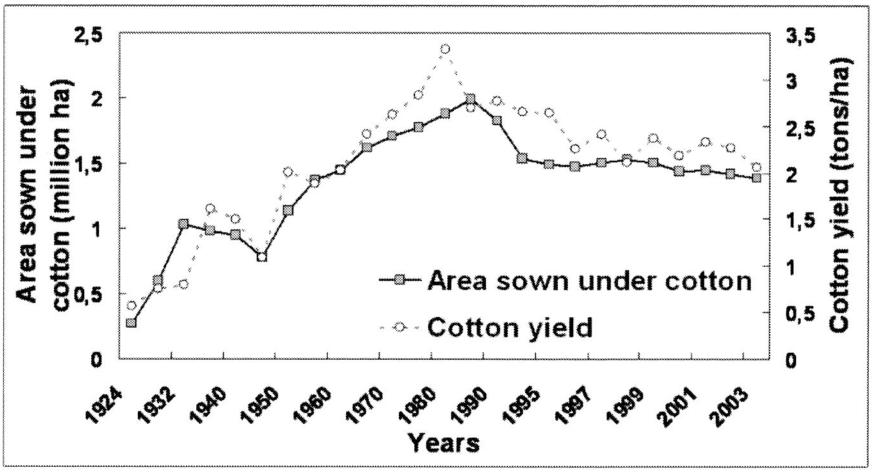

Figure 4. Area sown under cotton and its yield for the last century in Uzbekistan

The operation and maintenance of this complex hydraulic infrastructure was supported by the central Soviet administration. After the breakup of the Soviet Union, the financial flows have ceased substantially for rehabilitation works. For example, over a 10-year span (1991-2001) the government investment portfolio for irrigation infrastructure rehabilitation has decreased from 27% to 8% and the capital investment for construction to the water sector was reduced by 5 times (UNDP, 2007).

This reduction in the investment portfolio has further worsened the operation and maintenance of the existing infrastructure and everlasting increase of electricity prices. The increase for electricity costs have ramifications for the sustainability of irrigation infrastructure because as early mentioned about half of total irrigated lands (2.2 million ha) in Uzbekistan are operated via various lift and diversion facilities. Indeed about 80% of the total Ministry of Agriculture and Water Resources of the Republic of Uzbekistan budget is devoted to electricity costs for the operation of lift infrastructure.

2.4. Irrigated Land Quality Classification System

An irrigated land quality classification system in Uzbekistan is termed as *bonitet.*[1] Bonitet Score reflects the soils potential productivity (i.e. yield potential of cotton) based on inherent fertility and quality (Noble et al., 2005). It is expressed as a score on the scale of 1 to 100 (Table4). Assessment of irrigated land quality is conducted over every 10 year span in Uzbekistan.

This assessment system is based on the relationship between indicators and characteristics of soils such as the humus content, soil texture, salinity level and erosivity with its productivity. The generalized characteristics of the irrigated land quality is quantified by a relative quantity values-bonitet score (Talipov, 1992; Shadibaev, 2009). The bonitet score is conducted relatively to a cotton yield of the best soil quality in Uzbekistan, (e.g. cotton yield

[1] Bonitet is used by Government officials to classify land into classes based on their potential productivity and quality to set annual production targets for a farm.

of 0.04 tons per 1 bonitet score and wheat yield of 0.06 tons per 1 bonitet score). Thus, the classes 1-4 are lands that need amelioration and remedial actions for improving the soils.

Table 4. Land quality classification in Uzbekistan

Classes	Description	Bonitet Score
10	Best	91-100
9	Best	81-90
8	Good	71-80
7	Good	61-70
6	Average	51-60
5	Average	41-50
4	Below average	31-40
3	Below average	21-30
2	Bad	11-20
1	Bad	0-10

The recent official average value of the bonitet score for Uzbekistan is estimated to be 55. However the actual value is estimated to be 52-53. Figure5 depicts the evolution of the bonitet score for Uzbekistan. The bonitet score decreased from 60 in the 1970s to 52 in 2000s. As the principle of bonitet classification system was designed to estimate the cotton yield from the field plot. The bonitet score is still used a major indicator for potential crop yields of an agricultural land plot and thus lands with higher bonitet score are more attractive for farmers.

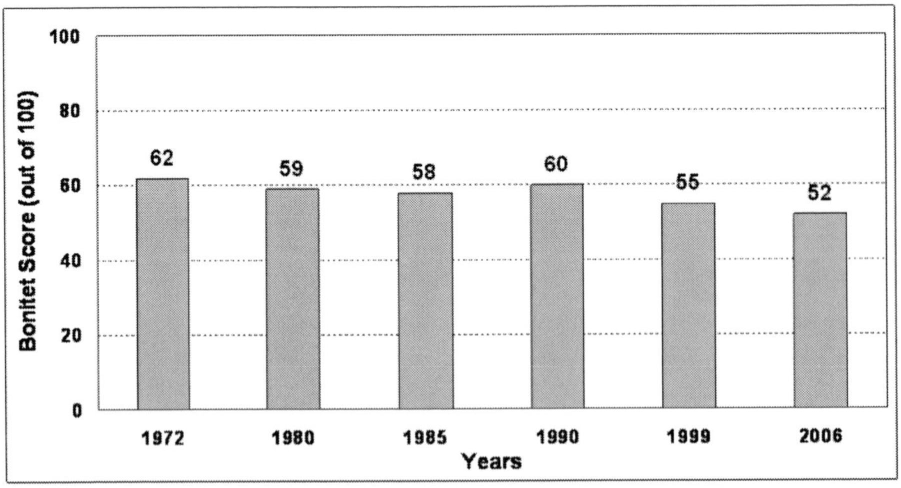

Figure 5. Institutional water management hierarchy in Uzbekistan

3. Environmental Threats

According to UNESCO (2000) the average percent of irrigated land affected by salinization in Central Asian region is about 23% with a range of in the low 6% in Kyrgyzstan and in the high 50% in Uzbekistan. In addition, the groundwater table has increased and the area of irrigated land with high groundwater levels (less than 2 meters below the surface) exceeds 30% of the total irrigated land in the region (Toderich et al., 2008).

Sustainability of irrigated agriculture in Uzbekistan is threatened by the salinization of land and water resources (Tookey, 2007). The land and water quality degradation is observed as a result of the excess water withdrawal over the actual crop requirements that impact the crop yields, decreased aggregate national income and affect the livelihoods of population (Murray-Rust et al., 2003).

These problems are the result of seepage from unlined canals, inadequate provision of surface and subsurface drainage and poor water management. Approximately, half of the irrigated area falls under different types of salt-affected soils and average yield of cotton losses may be as high as 50 percent. The annual economic losses due to land salinity and its abandonment are estimated to be more than US$43 million in Uzbekistan (World Bank, 2003; UNDP, 2007). Though the loss of crop production due to soil salinization is important, salinized land is generally still cultivated since no alternatives are available at the present time.

4. Institutional Management

There have been various main transformations in agricultural production system through primitive nomadic agricultural practices to the sophisticated and heavy mechanized agricultural production in the Central Asia region. Table5 depicts the major transformation phases in agricultural farming systems in Uzbekistan for the last century.

Table 5. Typology of farming systems in Uzbekistan

Period	Type of farming system	Ownership	Command area (ha)
Prior 1900’s	Nomadic	Private	various
1920-1940	Kolkhoz	State cooperative farm	10 000-20 000
1940-1980	Sovkhoz	State farm	Up to 100 000
1990-2006	Shirkat	Semi-Cooperative farm	1500 - 2000
2005-present	Fermer	Private farm	20
2000-present	Dekhkan	Individual farm	1

4.1. Ancient Period

The earliest material evidence of irrigated agriculture is dated to the sixth millennium BC, in the simplest form of liman (flood-land) farming (Mukhamedjanov, 1978). The evidence also demonstrates a shift from simple use of water resources to increased regulation of seasonal river flows though melioration of individual plots, diverting surplus water and cleaning silt from riverbeds. This allowed increasing the area under irrigation.

In the latter part of the VII century when the region came under Arab control, extensive and sophisticated irrigation networks were constructed which could deliver water for many kilometres from its source to the irrigated lands (Irrigation of Uzbekistan, 1979; Mukhamedjanov, 1978; O’Hara, 2000; Rakhmatullaev et al., 2003). In this period, the world famous remnants of civilization such as Samarkand, Bukhara, Khiva, and Kokand (Uzbekistan), Merv (Turkmenistan) have emerged.

There were three *khanates* (kingdoms) – Kokand, Khiva and Bukhara from the XIII to late the XIX centuries in the territory of modern Uzbekistan. Figure6 depicts the hierarchy over water management in ancient times in Uzbekistan.

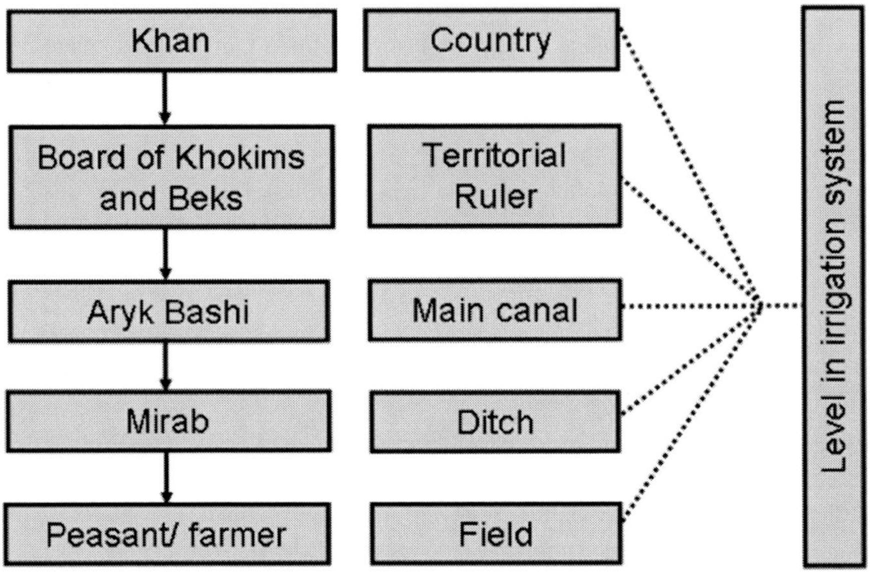

Figure 6. Hashar working for the cleaning of an irrigation canal in ancient Khorezm (Irrigation of Uzbekistan, 1979)

The *khan* (king) was in the head of the water delivery hierarchy. Each of them had a board of *khokim* and *bek*, local rulers who were appointed by the khan. Each administrative territory was managed by a khokim or bek. Each khokim or bek had appointed *aryk bashi* on the main irrigation canal. The aryk bashi is a person who reported on the water level in main irrigation canal and decided what water intake gates to be opened or closed (Mukhamedjanov, 1978; O'Hara, 2000). On the level of irrigation ditches, a *mirab* was responsible for water allocation and operation of irrigation ditches. The mirab was elected by local people in general and local farmers from respected persons within the community for water allocation and distribution. The *dekhkans,* peasant farmers, were the end users of the irrigation water (Rakhmatullaev et al., 2003).

The land belonged to the khan and the dekhkans had a long term rent for their plots. The size of the land rented to dekhkans depended on the family size. The tax or rent payment was in the form of raw materials, mainly quarter part of the grain harvest. The average assessment fee was 1/10 of the harvest and it usually went to the mirab. However, the payment was never a consistent percentage of the crop, as the farmer paid depending on how satisfied he was with the job that the mirab was doing (Irrigation of Uzbekistan, 1979; O'Hara, 2000). Then the mirab estimated the needed amount of water and paid for the water to the arik bashi. Therefore, the mirabs and dekhkans had no interest in taking any extra amount water that exceeded their needs (Muhamedjanov, 1978).

The dekhkans were free to move to other kingdoms. Therefore, more dekhkans in the kingdom was in the direct interest of each khan. In VI and VII centuries, the region experienced the economical growth in the routine of the Great Silk Road. The taxation system was organized in a very smart way, i.e., one-fourth of the harvest was taxed whatever the dekhkans grow (Rakhmatullaev et al., 2003). This system was successful because the dekhkans decided what to grow. Usually the agricultural production was market driven. Alternatively, whatever agricultural goods were expensive and needed in the local markets (bazaar), the favor was given to those crops in a given year. Therefore, the state's policy was dependent on market concepts.

For the maintenance and rehabilitation works of the existing irrigation canals and construction of new hydraulic infrastructure the *hashar* was practiced in the region (Figure 7). Hashar is the action that pulls community to accomplish maintenance, rehabilitation and construction works on a voluntary basis in Uzbekistan. This form of actions still widely practiced in Uzbekistan. The hashar accomplished two things. First, it acts as a mobilization mechanism for labour input and material resources within community. Second it was a mean of maintenance of the infrastructures (O'Hara, 2000; Rakhmatullaev et al., 2003).

Figure 7. Hashar work for cleaning irrigation canal in ancient Khorezm (Irrigation of Uzbekistan, 1979).

For example, for the construction or rehabilitation of irrigation canals, mirabs or aryk bashi called for hashar works. Each village had to mobilize their labor force for hashar. Villages at the upstream of the irrigation canal who would expect to receive more and cleaner water were expected to contribute more in terms of time and resources. Hashar was obligatory to all water users to take part in this annual maintenance works. Individuals who refused to participate were fined or denied access to water. Thus, "hashar was a system that linked benefits to duty" (O'Hara, 2000).

Prior to be part of the former Soviet Union, Uzbekistan came under Tsarist control after the invasion of Russia. With the new administration, irrigation reforms were attempted to be introduced. However, these reforms failed because of many appointed irrigation officials were unaccustomed to the traditional methods of management and the authorities declared that irrigation would be run "by custom" (Thurman, 1999).

4.2. Soviet Period

There have been two main types of farming systems in the former Soviet Union *kholkhoz* [2] (cooperative farm) and *sovkhoz* (state farm). The principal difference between these two forms of ownership was that a sovkhoz was a state enterprise whose workers were employed at fixed wages. By contrast, a kolkhoz paid its workers from its own annual earnings (Yalchin and Mollinga, 2007). According to Talipov (1992) by 1940 the collectivization of the former private farms (around 800,000 farms) into collective farming systems (kolkhoz) was accomplished through the establishment of 7629 kolkhoz. However, by the end 1990's the proportion of sovkhoz has been increased due to the fact that sovkhoz was considered as a more socialist production entity in contrast to the cooperative ownership of kolkhoz. As a result, by the end of 1990's only 9.5% and 63% of the total agricultural lands were managed by kolkhoz and sovkhoz respectively.

The kolkhoz type was mainly specialized in the production of cotton, fruits and vegetables, tobacco and other (lemon and fishery). On the other hand, the sovkhoz was specialized in the production of husbandry, poultry, vineyards, rice, wheat, cotton and others. In 1991, the total number of kolkhoz was 941 with a total irrigated area of 1.5 million ha whereas the total number of sovkhoz was 1017 with a total irrigated area of 2.1 million ha (Talipov, 1992).

In order to increase the agricultural production, the soviets implemented various progressive measures such as mechanization, agronomic practices, development of irrigation infrastructure. As all kolkhoz and sovkhoz were specialized in the production of certain agricultural product, a central management body compromising with trained irrigation engineers, agronomists, veterinarians and other professionals was put into place. Irrigation water distribution was carried out by professionals and much of subsidies were granted for collective farms. However, the processing and manufacture aspects of agricultural produce were neglected in favor of inter-republic trade in the soviet times (Yalchin and Mollinga, 2007).

4.3. Post-Independence Period

Since the independence of Uzbekistan the situation has changed dramatically in terms of institutional, political and technical systems (Veldwisch, 2007). Political transition from a planned economy to market has introduced "new" concepts like land tenure, water rights and

2 Kolkhoz (Russian). A large collective farm comprises several agricultural experts and farm laborers responsible for agricultural production and delivery targeted outputs for cotton and wheat to the government. A typical area under management of kolkhoz ranges in size from 10,000 to 20,000 ha (Noble et al., 2005).

different kinds of ownership (Morgounov and Zuidema, 2001; Yakubov and Ul Hassan, 2007).

In 1990, the former collective farms were transformed to collective family farming units called *shirkat*. The shirkat is a collective farm which pays a fixed monthly salary and supplementary dividends from its annual profits for its workers (Yalchin and Mollinga, 2007). The organizational management of the shirkat was similar to the former kolkhoz system with significant difference in contract arrangements for crop production. The shirkat has a single management unit comprising of a head, professionals (engineer, agronomist, accountant and etc). The shirkat contracts several brigades (working units) for agricultural production on annual basis for meeting state quota for cotton and wheat yields. In addition, every year the workers would sign a contract with the shirkat management for leasing a plot with his/her state quota for cotton and wheat.

As the shirkat farming system has shown its inefficiency, private farm system (*fermer* and *dekhkan*) has been advocated. The main difference between the two types of farming systems is that fermer farms have contract with the central government to meet the targeted cotton and wheat production whereas dekhkan farms do not have such contracts for meeting the targeted cotton and wheat crop production and only produce for meeting their own needs. This is true for only cotton and wheat productions (Yakubov and Mathrithilake, 2009).

Fermer is a private farm, successor of kolkhoz and responsible for the targeted production of cotton and wheat to the government. The fermer can lease the plot in size ranging from a minimum of 10 ha to several hundred ha, with an average size of approximately 20 ha (as of 2000). These fermer farms have flexibility in the hiring of labor and access to other subsidized inputs from government (such as fertilizers, lease of machinery and access to bank credits). According to the Ministry of Agriculture and Water Resources of the Republic of Uzbekistan (2009) as of 2009 there are about 220,000 fermer farms in Uzbekistan.

Dekhkan farmers have an average plot size of 0.12 ha and do not have the right to hire workers: only family members can work on the plots, which are given to the dekhkan farmers on a lifelong lease and can be passed on as inheritance to their children. There are more than 3.5 million dekhkans in Uzbekistan (UNDP, 2007). Those who are not employed by the new farms (fermer) are officially classified as dekhkan farmers, and work on the small plots allocated to them in the process of land distribution.

All agricultural land is still technically owned by the state and is not tradable or transferable. Leases of private farms (fermer and dekhkan) are in principle valid for a minimum of 10 and a maximum of 50 years. Fermer farms have often been subject to arbitrary decisions on the part of local authorities. For example, their land plots can be taken away and reallocated if they do not comply with local authorities' decisions about its use or have shown poor performance over targeted production of cotton and wheat.

The further reforms have been initiated in agricultural production system such as enlargement of fermer farm units in 2008 in Uzbekistan (MAWR of Uzbekistan, 2009). The main objective of recent reforms is to consolidate smaller poor performing fermer farms into efficient large fermer farms. Before this reform there were about 220 000 fermer farms in Uzbekistan with an average agricultural land of 27 ha per fermer farm whereas after the consolidation process the average size of fermer farm becomes 57 ha totaling of 104 000 fermer farms (MAWR Uzbekistan, 2009).

5. Current Challenges and Perspectives

The major current challenge facing irrigated agriculture in Central Asia and specifically in Uzbekistan is weak water management institutions, i.e., water users associations at the on-farm level. It is known that water management is a political process and important part of overall development policy of a given country (Abdullaev et al., 2009). Many international donor assistance programs have been launched and implemented for institutional interventions in water management in the region. However, interrelated aspects such as financial, technical and governance issues have been bottlenecked for proper progress of such interventions. Prior to discussion of WUA challenges first the farmers current challenges will be discussed in regard to irrigated agriculture.

5.1. Challenges for Farmers

Land Tenure

As previously stated the agricultural lands are not privatized and technically owned by government in Uzbekistan. The farmers are interested in investing their resources for the amelioration of lands and improving the irrigation infrastructures in order to increase the agricultural production and their incomes. However, there is no guarantee that their lands can not be taken out and re-allocated to different farmers on arbitrary decisions of the local authorities. As a result, farmers fear to lose their invested capitals. In Uzbekistan, under the following conditions the farmer land can be taken away and re-allocated: 1) not meeting the targeted cotton and wheat state quota; 2) bankruptcy of farmers; 3) non transparent administrative procedures.

In addition, the farmers fear to invest their own financial resources because the re-allocation can only be carried out by state authorities and farmers cannot sell their land leases to other farmers in order to partially compensate already invested funds. In contracts between the state and farmers there is no clear identification of mutual rights for co-financing of irrigation infrastructures rehabilitation works. In addition, there are no incentives (reduction of state quota for cotton and wheat, tax returns) for farmers in cases of investing resources for rehabilitation of irrigation infrastructures.

Cropping Structure

The state quota for cotton and wheat is based on the bonitet scoring system in Uzbekistan as previously discussed. The main bottlenecked aspect is that the bonitet score was carried out in late 2000s and is now outdated. Most probably the current scores are higher because in most instances the ameliorative soil conditions have deteriorated to date and farmers cannot meet their targeted state quota which in turn is putting them into the high risk of losing their lands. For example, in a given land plot, previously prescribed bonitet score is used for the estimation of cotton and wheat yields. Local authorities are reluctant to the reduction of area sown under cotton and wheat, although farmer can meet the targeted yield on a smaller area with new fertilizer application or other innovative techniques. The local authorities fear that the decrease of cotton and wheat sown areas can be viewed from higher management hierarchy as non compliance to the common rules. In fact, there is a perception that by reducing the overall area prescribed for sowing cotton and wheat would result a lower crop harvest.

Consulting/Extension Service

After the fundamental transformation processes in Uzbekistan, the farmers are left without strong extension/consulting services either private or quasi-governmental. Many farmers are willing to pay for consultancies on new irrigation methods and technologies, cost-effective amelioration techniques, water conservation, legal aspects and marketing. However, there is no legislation for government or private extension and consultancies services in Uzbekistan. Thus the farmers are striving to overcome their current problems with old fashioned common known irrigation and agronomic techniques.

5.2. Challenges for Water Users Associations

A WUA is a self-managing group of farmers working together to operate and maintain their irrigation and drainage network (only inter-farm or on-farm level) to ensure fair and equitable water distribution and increase crop yields (Kazbekov et al., 2009). The WUA was established to replace the shirkat for on-farm irrigation infrastructure (canal and small water diversion facilities with gates) operation, maintenance and water distribution. According to the official statistics the total number of WUA is 1711 in Uzbekistan (MAWR of Uzbekistan, 2009). WUAs already serve around 3.6 million ha of land and 111,000 km of irrigation canals in Uzbekistan.

Most of these WUAs were established in the 2003–2006 period and now play an important role in the allocation and distribution of irrigation water, the collection of irrigation service fees and the maintenance of the irrigation infrastructure (Abdullaev et al., 2008). However due to the poor technical and financial capacities of farmer members of these institutions (WUA) and WUA itself is very limited institution for proper operation and maintenance of irrigation infrastructures.

Territorial vs. Hydrographic

The current WUA is established on the former kolkhoz territory, i.e., most of the WUA are organized on administrative territorial principle not the hydrographic. A farmer signs an annual contract for the irrigation water delivery with WUA in form of its annual membership fee. The geographic location of farmer's field parcel can force him/her to sign several contracts with different WUAs either because of administrative boundaries or several irrigation canals pass through that farmer's land plot. There are also positive aspects of such geographic location such as a guaranteed water supply but overall the farmer overpays his dues in comparison with the rest within the concerned area. The most complication arises for farmers located at the tail-end of the irrigation system, especially in dry years.

Financial Incapacity

Thus the WUA barely can support its limited personnel with monthly salaries from collecting membership fees from individual farmers. In addition, the most professional and trained specialists once worked at collective farms have left for other areas of employment such as business and trade. The current WUA does not have in terms of quantity and quality experts for consulting local farmers on irrigation regimes, cropping structure, appropriate water and agronomic practices and business planning.

WUA's budgetary resources are based on the individual farmers' membership fees. At the moment, each farmer signs an annual contract on irrigation water delivery with the WUA. However, the farmers cannot pay on a timely manner their respective membership fees to WUA because many of them can receive their earned money for cotton and wheat (state quota ordered crops) only from the previous harvest year. The state quota system on cotton and wheat has been designed in such a way that farmers on its obligation for meeting targeted harvest should produce and sell at government fixed prices. On the other hand, the government subsidizes the inputs (fertilizers, water) for farmers. The farmers can use the rest of low-cost fertilizers for producing other cash crops, fodder for husbandry practice and partially compensate their income earnings. The state quota system is put under critics by many scholars and experts for being an inefficient system for agricultural production.

Under Uzbekistan present conditions, the main detrimental factor of this system is not the monetary value of the purchased price of cotton and wheat by government but rather the complicated control of fiscal and financial system, i.e. substantial delays of getting money from banks. In addition, the farmers cannot use their earned funds for barter arrangements.

Increase Number of Farmers

With the de-collectivization of former soviet collective farms and the emergence of new private forms of farming systems there has been a rapid increase of the number of farms specialized in different cropping structures along a single irrigation canal. Thus each cropping system requires different irrigation application regimes. Many farmers favor to use only simple flooding irrigation methods instead of costly drip irrigation and other alternatives. Thus, WUA can hardly cope with the large number of farmers.

Technical Aspects

The most difficult aspect is the technical deterioration of existing irrigation infrastructures and improper allocation of funds for operation and maintenance works. There are widely non-existing or heavily deteriorated water measuring structures in place and thus it is extremely inconvenient for monitoring the right amounts of water. In fact, only nominal water measuring mechanisms are widely carried out for payments option of irrigation delivery service fees.

5.3. Perspectives

Farmers

The most practical and viable recommendations for individual farmers should be the regulation of legal arrangements defining the concrete and transparent mechanisms for the re-allocation procedure of land tenure rights from local government authorities. The most important is that the farmer should have a right to sell his/her land lease for another potential farmer in order to compensate his/her invested assets. Then farmers would be on the safe side and motivated to invest their own technical and financial resources for amelioration of lands and for rehabilitation and maintenance of irrigation infrastructures.

The second paramount aspect is the concrete definition of responsibilities and rights of farmers and government for the amelioration of lands and rehabilitation of irrigation infrastructures in land lease contracts signed between the farmer and the government. In most cases, there is no clear definition of such rights or whole responsibility shifted to farmers. As a result, farmers are at high legal risks under court rulings. The clear and concrete definition of such responsibilities and rights would stimulate farmers to improve the amelioration conditions of their lands and actively participate with technical and material contributions for rehabilitation of irrigation infrastructures.

The third aspect is the incentive mechanisms for farmers' contributions in such activities. For example, the decrease of state quota for cotton and wheat either in quantity or options for farmers to sow the two crops on less parcels while meeting the state quota.

The fourth aspect is that the current bonitet scores should be carried out for all lands and updated accordingly by the responsible government authorities. In addition, each farmer can hire licensed private consulting company for independent bonitet scoring. There is a dilemma as the bonitet score can only be carried out by specialized state organization, thus such licensing should be shifted to private sector and necessary amendments into the regulations should be envisaged. Proper and updated bonitet score would be beneficial both for farmers and government in order to meet state quota on cotton and wheat.

The fifth aspect is provision of farmers with a more flexible system for cultivating different cash crop varieties on their plots. If the government would allow the decreasing of the size of lands sown under cotton and wheat while meeting the state quotas, the remaining lands could be used effectively for cultivation of cash crops. As a result, the income generation would be promising and farmers could contribute for amelioration works and rehabilitation of irrigation infrastructures. There was a promising sign from the government of Uzbekistan in 2008, who has issued the degree that the total size of area sown under cotton should be reduced by 75 000 ha and these lands can be used for cultivation of different crops (Shadibaev, 2009).

Water Users Associations

The first aspect is the development of mechanisms for financial sustainability of WUA. The only source of income for WUA is the annual membership fees from its farmer members. Instead for diversifying the income sources in kind contributions should be introduced as in Kyrgyzstan and Tajikistan (Sehring, 2009). In kinds contributions can be either natural produce or manpower for cleaning irrigation infrastructures. For example, the farmer who installs on his/her own expense the water measuring device can have some privileges, i.e. first irrigation water delivery.

The next aspect is consultancy/extension service for farmers. As Muhamedjanov et al (2008) proposes that consultancy/extension services should be established under the umbrella of the of Basin Irrigation Systems Authority (BISA) which were established for replacing previous provincial water management organizations in Uzbekistan. The main disadvantage of such consultancy/extension service is emergence of another inefficient and bureaucratic machine. Instead the consultancy/extension should be in solely form of private companies at the WUA level. The reasons are that WUA is directly working with individual farmers and such private companies with WUA partnership will have substantial impacts as cost-effectiveness and targeted consultancy service provision. Another important aspect as early mentioned is that there is no legal framework for such consultancies. For extension service the prominent academic universities and research institutions with WUA and private consultancy companies can have synergetic partnerships for training and know-how technology transfer.

Conclusion

The Central Asian countries have faced new challenges for practicing a sustainable irrigated agriculture, particularly in terms of institutional challenges. The physical deterioration of irrigation and drainage infrastructures is a technical issue that would impose constraints on agricultural production, environmental concerns and, as a result, on the livelihoods of many millions of farmers. In fact poor conditions of irrigation infrastructures impact the efficient irrigation water distribution at various levels of the irrigation conveyance system. Global climate change and the transboundary nature of water resources would be major supplemental aspects for the sustainability of irrigated agriculture production in Central Asia.

The national and international donor programs would play a pivotal role for reversing the current situation if the intervention measures would be targeted to institutional changes such as land tenure, water rights, and incentives for rehabilitation of irrigation and drainage networks.

Acknowledgments

The authors would like to thank the French Embassy in Tashkent for supporting the French-Uzbek cooperation in the field of agriculture and water sciences. This study has also been supported by INTAS fellowship Nr. 04-83-3665 and by the French Ministry of Foreign Affairs via the Eiffel fellowship program No. 530909C. However, the views expressed in this paper do not necessarily reflect those of the funding agencies, and no official endorsement should be inferred from it.

References

Abdullaev, I., Kazbekov, J., Manthritilake, H., & Jumaboev, K. (2009). Participatory water management at the main canal: A case from South Ferghana canal in Uzbekistan. *Agricultural water management, 96* (2), 317 – 329.

Abdullayev, I., Nurmetova, F., Abdullaeva, F., & Lamers, J. (2008). Socio-technical aspects of Water Management in Uzbekistan: emerging water governance issues at the grass root level. In M. M. Rahaman, O. Varis (Eds.), *Central Asian Waters: Social, economic, environmental and governance puzzle*. 89-103. Helsinki, Finland: Helsinki University of Technology.

Dukhovny, V. A. (2007). Water and globalization: Case study of Central Asia. *Irrig and Drain, 56*, 489-507. DOI: 10.1002/ird.327.

FAO (Food and Agriculture Organization of the United Nations). (2009) *http://www.fao.org*. [Accessed on August 2009].

Green, D. J. (2001). Regional co-operation policies in Central Asia. *Journal of International Development, 13*, 1151-1164.

Goskompriroda of Uzbekistan (State Committee for Nature Protection of the Republic of Uzbekistan). (2005). Conditions of environment and use of natural resources in Uzbekistan (2002-2004). National report. Tashkent, Uzbekistan.

Hagg, W., Braun, L. N., Kuhn, M., & Nesgaard, T. I. (2007). Modelling of hydrological response to climate change in glacierized Central Asian catchments. *Journal of Hydrology, 332*, 40-53.

IAMO (Leibniz Institute of Agricultural Development in Central and Eastern Europe). (2008). Continuity and change: Land and water use reforms in rural Uzbekistan: Socio-economic and legal analyses for the region Khorezm. Vol.43, *http://www.aimo.de/dok/sr_vol43.pdf*.

Irrigation of Uzbekistan: Current condition and perspectives of irrigation development in Amu Darya river basin. (1979). Volume III. Tashkent, Uzbekistan: Publisher "Fan".

Ibragimov, N., Evett, S. R., Esanbekov, Y., Kamilov, B. S., Mirzaev, L., & Lamers, J. P. A. (2007). Water use efficiency of irrigated cotton in Uzbekistan under drip and furrow irrigation. *Agricultural Water Management, 90*, 112-120.

Kandiyoti, D. (1999). Rural Domestic Economy and Female Labour Supply in Uzbekistan: Assessing the Feasibility of Gender Targeted Micro Credit Schemes. Final Report for Department of International Development, ESCOR Unit. Grant No: R6978.

Kazbekov, J., Abdullaev, I., Manthrithilake, H., Qureshi, A., & Jumaboev, K. (2009). Evaluating planning and delivery performance of Water User Associations (WUAs) in Osh Province, Kyrgyzstan. *Agric. Water Manage*. DOI:10.1016/j.agwat.2009.04.002.

Lioubimtseva, E., Cole, R., Adams, J. M., & Kapustin, G. (2005). Impacts of climate and land-cover changes in arid lands of Central Asia. *Journal of Arid Environments, 62*, 285-308.

MAWR Uzbekistan (Ministry of Agriculture and Water Resources of the Republic of Uzbekistan). (2009). *http://www.agro.uz* [Accessed 5 July 2009].

Morgounov, A., & Zuidema, L. (2001). The Legacy of the Soviet agricultural research system for the Republics of Central Asia and the Caucasus. Research Report No. 20. The Hague: International Service for National Agricultural Research.

Mukhamedjanov, Sh. Sh., Galustyan, A.G., Nerozin, A.S. (2008). The end water user and consulting service for farmers. In V.A. Dukhovny, V.I. Sokolov, H. Manthrithilake (Eds), Integrated water resources management: from theory to practice. Experience from Central Asia. 140-157. Tashkent, Uzbekistan: Scientific Information Center of Interstate Commission for Water Coordination.

Mukhamedjanov, A. R. (1978). *History of irrigation in Bukhara oasis (from ancient to the beginning of XX century)*. Tashkent, Uzbekistan: Publisher "Fan".

Murray-Rust, H., Abdullaev, I., Ul Hassan, M., & Horinkova, V. (2003). Water productivity in the Syr-Darya river basin. Research Report 67. Sri Lanka: International Water Management Institute.

Nezlin, N. P., Kostianoy, A. G., & Lebedev, S. A. (2004). Interannual variations of the discharge of Amu Darya and Syr Darya estimated from global atmospheric precipitation. *Journal of Marine Systems*, *47*, 67-75. DOI:10.1016/j.jmarsys.2003.12.009.

Noble, A., Ul Hassan, M., & Kazbekov, J. (2005). "Bright spots" in Uzbekistan, reversing land and water degradation while improving livelihoods. Research Report 88. Sri Lanka: International Water Management Institute.

O'Hara, S. L. (2000). Lessons from the past: water management in Central Asia. *Water Policy*, *2*, 365-384.

Rakhmatullaev, S., Huneau, F., Kazbekov, J., Le Coustumer, P., Jumanov, J., Motelica-Heino, M. & Hrkal, Z. (2009). Groundwater resources and management in Amu Darya river basin (Central Asia). *Environmental Earth Sciences*. DOI:10.1007/s12665-009-0107-4.

Rakhmatullaev, S. A., Bazarov, D. R., & Kazbekov, J. S. (2003). Historical irrigation development in Uzbekistan from ancient to present: Past lessons and future perspectives for sustainable development. In Proceedings of the third International Conference of International Water History Association. Alexandria, Egypt, pp.79-80.

Royal Haskoning (2001). Aral Sea basin program. Water and Environmental Management Project. Sub-component A1: Salt and water resources management on regional and national levels. Regional Report No.2. Report on Phase III- Regional needs and limitations. Technical report. Global Environmental Facility and International Fund for Saving the Aral Sea. 15 November 2001

Sattarov, M. A., Eshmirzoyev, I. E., & Rakhimov, F. (2006). Problems of estimation and rational use of water resources of Aral Sea basin. Proceedings of International Conference on Extreme Hydrological Events in Aral and Caspian Sea Region. Moscow, Russia, pp. 269-274.

Sehring, J. (2008). The pitfalls of irrigation water pricing in Kyrgyzstan and Tajikistan. *Development*, *51*, 130-134. DOI:10.1057/palgrave.development.1100452.

Shadibaev, T. (2009). The lands can be productive, with special treatment and management. *J. Economic Review*, 3(*114*). 12-17.

Shultz, V. (1949). *Central Asian rivers*. Moscow, Russia: Publisher "Nauka".

Sokolov, V. (2006). Experiences in IWRM in the Central Asia and Caucasus Region. *Water International, 31*(1), 59–70.

Statistics Committee of Uzbekistan (State Committee of the Republic of Uzbekistan on Statistics). (2006). Environmental protection in Uzbekistan: Statistical bulletin. Tashkent, Uzbekistan: State Committee of the Republic of Uzbekistan on Statistics.

Talipov, G. A. (1992). *Land resources of Uzbekistan and their rational use*. Tashkent, Uzbekistan: Publisher "Uzinfopmagroprom".

Toderich, C., Tsukatani, T., Shoaib, I., Massino, I., Wilhelm, M., Yusupov, S., Kuliev, T., & Ruziev, S. (2008). Extent of salt affected land in Central Asia: Biosaline agriculture and utilization of the salt-affected resources. Kier Discussion Paper Series. Discussion Paper No. 648: Kyoto of Economic Research: Kyoto University, Kyoto, Japan. *http://www. kier. kyoto-u.ac.jp/index.html*

Tookey, D. L. (2007). The environment, security and regional cooperation in Central Asia. *Journal of Communist and Post-Communist Studies, 40*, 191-208.

Thurman, M. (1999) Modes of organization in Central Asian irrigation: The Ferghana valley, 1876-present. PhD thesis. University of Indiana, Indiana, USA.

UNDP (United Nations Development Programme). (2009). UNDP: Uzbekistan in Figures. *http://www.statistics.uz/data_finder/74/*. [Accessed 20 August 2009].

UNDP (United Nations Development Programme). (2007) *Water, critical resource for Uzbekistan's future*. Tashkent, Uzbekistan: UNDP-Uzbekistan.

UNEP (United Nations Environmental Programme). (2005) *Global inter national waters assessment Aral Sea. GIWA Regional assessment 24*. Kalmar, Sweden: University of Kalmar.

UNESCO (United Nations Educational, Scientific and Cultural Organization). (2000) *Water-related vision for the Aral sea basin for the year 2025*. Paris, France: UNESCO.

Veldwisch, G. J. A. (2007). Changing patterns of water distribution under the influence of land reforms and simultaneous WUA establishment: Two cases from Khorezm, Uzbekistan. *Irrigation Drainage Systems*. DOI: 10.1007/s10795-007-9017-3.

World Bank. (2005). *Drought: Management and mitigation assessment (for Central Asia and the Caucasus*). Report No.31998-ECA.

World Bank. (2003). *Irrigation in Central Asia: Social, economic and environmental considerations*. Report. *www.worldbank.org/ eca/ environ ment*. [Accessed on 12 January 2008].

Ximing , C., McKinney, D. C., & Rosegrant, M. W. (2000). Sustainability analysis for irrigation water management in the Aral Sea region. *Journal of Agricultural Systems*, *76*, 1043-1066.

Yakubov, M., & Manthrithilake, H. (2009). Water for food as food for thought: Case study of applying the podiumsim model to Uzbekistan. *Irrigation and Drainage*, *58*, 17-37.

Yakubov, M., & Ul Hassan, M. (2007). Mainstreaming rural poor in water resources management: Preliminary lessons of a bottom-up WUA development approach in Central Asia. *Irrigation and Drainage, 56*, 261–276. DOI: 10.1002/ird.293.

Yalcin, R., & Mollinga, P.P. (2007). Institutional transformation in Uzbekistan's agricultural and water resources administration: *The creation of a new bureaucracy. Working Paper Series* No. 22. ISSN 1864-6638. Bonn, Germany: Center for Development Research (ZEF) University of Bonn.

In: Agricultural Production
Editors: Felix C. Wager, pp. 163-179
ISBN: 978-1-61668-695-6

Chapter 8

AGRICULTURAL PRODUCTION IN CHINA

Jenifer Huang McBeath* and Jerry McBeath≠
University of Alaska Fairbanks

ABSTRACT

In the last two decades, the Chinese government has devoted large resources to insure that agricultural production keeps pace with a growing population and changing environment. This chapter focuses on seven areas influencing food production: 1) maintenance and expansion of the amount of arable land; 2) improvement and expansion of the irrigation system; 3) afforestation/reforestation, primarily to counter erosion and flooding; 4) creation of transgenic seeds of food crops that are pest- and drought-resistant; 5) other germ plasm improvements such as the development of hybrid varieties of rice and improvements of other varieties; 6) control of plant diseases and insect pests through cultural, chemical, and biological measures; 7) improvements of cultural (farming) practices, such as multiple and inter-cropping, use of wind tunnels and mulching. Analysis of these efforts suggests that insufficient attention has been paid to the areas most likely to significantly improve productivity, namely plant disease/pest control and application of improved cultural practices.

INTRODUCTION

In 2010, China is the world's largest producer of agricultural goods. This would seem unremarkable in one sense, because China also is the world's most populous nation, with more than 1.3 billion residents and 200 million farms. However, only about 13 percent of China's landmass (globally the third largest) is suitable for agricultural production, and that is considerably below the international average on a per capita basis. Over the millennia, the Chinese people have struggled to be self-sufficient in food production and in most eras have succeeded.

---

* jhmcbeath@alaska.edu
≠ gamcbeath@alaska.edu

The current threats to agricultural production in China may be more serious than those previously encountered; they include population pressures, urbanization, economic development and resulting environmental crises. Throughout most of its history, China was the most populous nation in the world, but population pressures increased geometrically after the establishment of the People's Republic of China in 1949 (Shapiro, 2001; Greenhalgh & Winckler, 2005). In the 50-year period from 1949 to 1999, population doubled, from 542 million to 1.2 billion, but after the adoption of the "one-child" (with exceptions) policy in 1979, growth slowed. Still, population is expected to increase to 1.6 billion by the middle of the twenty-first century.

Population growth has brought corresponding increases in use of land for housing and human settlements, swallowing up huge stretches of agricultural land. Although the rural population has peaked at about 750 million, the urban population, greater than 500 million, continues to grow. Cities expand into the countryside and consume land once used for agricultural purposes (Di, 2004; Moody & Lan, 2010, 8). Increasing wealth and leisure of the growing middle class has led to conversion of rural land into golf courses, parks and other recreational uses.

Market reforms unfolding in China since 1978 have spurred economic development throughout the nation. Now, China is the world's third largest economic power and soon is expected to surpass Japan, the second largest economy. Increases in gross domestic product (GDP) of nearly 10 percent annually for many of the last 30 years are an historical record. Yet it is this rapid and essentially unplanned economic development that has seriously threatened agricultural land.

The environmental cost of rapid development is vast, and the literature of environmental studies is replete with examples of China's environmental crises (for early studies, see Ross, 1988 and Smil, 1993). They include degradation of the land through erosion, deforestation, desertification, and pollution; air pollution; degradation of the waters from chemical and other industrial effluents, excessive use of fertilizers and pesticides, and dumping of human wastes; and pollution of China's coastline and offshore ocean waters (McBeath & McBeath, 2009).

Our purpose in this chapter is to assess responses of the Chinese state to the aforementioned threats to agricultural production. We treat seven topics: restrictions on conversions of arable land into other uses; improvements in irrigation systems; control of erosion, desertification and flooding through re- and afforestation programs; development of agricultural biotechnology; improvements in native germ plasm; control of plant diseases and insect pests; and improvement of cultural (farming) practices.

The definition of food has changed, as China has developed economically and modernized. The diet of Chinese now includes a larger quotient of dairy products, fruits, vegetables, and meat, but space does not permit us to examine these areas of agricultural production. Instead, we focus on the traditional staples of the Chinese diet, often called the "five grains": rice, wheat, corn, millet, and potatoes.[1] Cultivation of each crop requires arable land, and for this reason we begin our analysis by discussing arable land loss.

1 Obviously, potatoes are not a grain crop; however, for statistical purposes in China, they are treated as though they were.

ARABLE LAND LOSS AND AGRICULTURAL PRODUCTION

Although China has a land area of 960 million hectares, its cropland is small—just 13 percent of the landmass (grasslands cover 42 percent; forests some 27 percent; and 18 percent is desert/other non-productive terrain). For the size of the population, this area of cultivation is small; in addition, soils are old and of less productive value than those of new areas as in Canada and the American Mid-West.

In 1995, Lester Brown (head of World Watch) predicted that China would need to import 200 million tons of grain by 2020 because of the loss of arable land (Brown, 1995), and that this large-scale entry into the global grain market would disrupt prices. This challenge to grain sufficiency occasioned a debate about the exact expanse of cultivatable land in China. Many experts, including Vaclav Smil (a geographer at the University of Manitoba who had extensive experience in land measurement of China), officials in the Ministry of Land and Resources and other Chinese experts joined the fray. Estimates of arable land range from Smil's optimistic count of 153 million hectares (Smil, 1999, 423-4) to the state estimate of 127 million hectares in 2007.[2] In our research in 2008-09, officials most often used the figure of 121.5 hectares, which corresponds with Premier Wen Jiabao's statement that China's arable land must not drop below 120 million hectares. Although there is no consensus on exactly how many hectares of arable land China has in 2010, it is clear that lands once used to cultivate staples have been lost.

From the 1980s, the central government used an hierarchical system to regulate the conversion of agricultural land to other purposes, but this system did not curb the incentives for local officials to bend rules to allocate collective land to commercial and industrial purposes, thereby increasing local revenues. In 2008, the central government revised the Land Management Law to re-centralize decision-making (Ho & Lin, 2004, 778). This revision, however, did not quell the strong pressure to convert agricultural land noted at the end of the Tenth Five-Year Plan in 2005. The regime then responded in three ways.

First, at the 17th Congress of the Chinese Communist Party in October 2007, Premier Wen Jiabao announced that China could not have less than 120 million hectares available for agricultural production. Reinforcing the premier's statement, the Minister of Land and Resources, Xu Shaoshi stated: "The red line of 120 million hectares of arable land cannot be crossed" (Zhao, 2007, 1). Second, the state established a new classification system for lands, barring construction of villas, golf courses, and other non-productive uses (Li, 2007, 3). The ministry tightened the land approval process and increased fees and penalties for illegal conversions (Li, 2007, 1). Third, China's land authorities made efforts to increase the amount of arable land (adding 2.4 million hectares from 1999 to 2006 [Xinhua, 2007, 3]), and began the process of consolidating land into larger parcels that could be farmed with a greater economy-of-scale (Xie, 2007, 2).

It appeared that the state, through its authoritarian powers, had stopped the loss of arable land, but later events suggest that this may not be the case. Responding to the global economic crisis of 2008-09, the central government inaugurated a $586 billion stimulus plan, and virtually every part of this plan required acquisition of land (Wang, 2008, 3).

2 This estimate of the National Bureau of Statistics is based on the area of cultivated land surveyed in late 1996, with revisions in 2001 (National Bureau of Statistics, 2007).

Simultaneously, responding to pressures to improve property rights in land for farmers, at the plenary session of the CCP Central Committee in October 2008, party leaders agreed to allow farmers to transfer their land use rights (Xinhua, 2008, 1). Although policy language paid lip service to "the most stringent farmland protection system (Wong, 2008, A1), the nature of policy change endorsed privatization of both state and collective land. Meanwhile, farmers increased their protests over loss of collective land to commercial use (Xinhua, 2008, A1).

To the present, the amount of arable land in China remains an important issue. We are not sanguine that the regime will be able to increase areas of cultivation, given enormous pressures to use scarce land for non-agricultural purposes. As we note below, however, the amount of land available for cultivation is not the primary factor influencing agricultural production today.

Irrigation and Agricultural Production

Lester Brown connected the nature of China's water system to agricultural production when, in 1998, he opined that depletion of aquifers and rivers running dry could dramatically increase China's need for grain imports (Brown, 1998). Although China has the fifth largest supply of water globally, on a per capita basis it is one-quarter below the global average. Estimates suggest that by 2030, per capita supply will drop to a level meeting the World Bank's definition of a water-scarce country (Turner & Otsuka, 2006, 1).

Currently, China's water resources are sufficient, but water is not evenly distributed throughout the country. Water is abundant in the humid south but relatively scarce in the north and west. Water use is inefficient; increasingly, as noted briefly above, water is polluted (for example, 16 of the world's 20 most polluted rivers are in China). Our concern is with the use of water in agriculture, which increasingly competes with industrial and personal/household use.

The state has made huge investments in irrigation infrastructure to maintain and enhance agricultural productivity. In the 1960s and 1970s, state spending on irrigation was 30 percent of total expenditures in rural China. Indeed, the state invested ten times as much in irrigation as it did in agricultural research (Fan et al., 2004, 417). More than one-half of the agricultural land in China is irrigated; surface water supplies two-thirds of irrigated areas, while groundwater supplies the rest (National Statistical Bureau, 2007). Among others, Huang et al. describe the "strong and robust" effect of irrigation on agricultural production (Huang et al., 2006, 30-52).

The Chinese state system is enamored of large-scale public projects, and water control developments are at the top of priority lists. For example, from 1949 to 1999, governments built 85,000 dams of different sizes (Ma, 2004, ix). The most controversial recent dam construction was at the Three Gorges of the Yangtze River, with an estimated overall cost of $37 billion. It aroused strong international (and some domestic) criticism because of the large number of people displaced by the construction, destruction of national cultural treasures, changes in regional climates (and increased flooding), and ecological destruction, among other complaints (Yardley, 2007).

The latest and largest project to improve irrigation in China is the South-North Water Diversion Project (SNWDP), which would address water scarcity in northern and western regions by moving water there from the south. SNWDP includes three water diversion routes—east, central, and west; it will link the Yangtze River of central China to the Huai, Yellow and Hai Rivers of the north (State Council, 2004, 15). Construction of the eastern route began in 2002 and is now scheduled for completion in 2013; it is the easiest to build as it can use existing rivers and lakes, including the Grand Canal and parallel rivers. The cost for this route alone is expected to be at least $7 billion, and the cost of the entire project, if completed, would exceed $75 billion in 2010 dollars (Economy, 2004, 126).

The first route was to have been completed by 2008, but serious water pollution conditions delayed work and increased costs. As a Huai River protection officer commented, "The common problem with China's pollution control initiatives is there are always loopholes in project implementation despite strict laws and rules" (Qian, 2010, 7). The second route is longer and will require new construction. The western route is still being designed and is the most controversial. It would divert Qinghai-Tibet plateau water resources from the Mekong and other international rivers, and the proposal already has provoked adverse reactions from China's neighbors in the southwest.

These controversies notwithstanding, the Chinese government definitely has taken seriously the need to maintain and expand irrigation projects for the purpose of securing continued high levels of agricultural production.

Control of Erosion, Desertification and Flooding Through Re- and Afforestation

A third area of state policy has been in response to extensive erosion, deforestation and desertification, which threaten arable land and make millions of hectares susceptible to flooding. Erosion of soils occurs naturally in ecosystems, but in China it mainly is a result of serious ecological destruction. A three-year survey released by the Ministry of Water Resources in late 2008 reported that China now has more than 3.5 million square kilometers of eroded land (Yu, 2008, 2), and erosion threatens more than one-third of China's territory.

About half of China's native forests have been destroyed since 1949, and today they cover little more than 14 percent of the land area. In addition to devastation of the Great Leap Forward and Cultural Revolution, mining and logging have deforested mountains, causing erosion, reduced water storage capabilities (including desertification), and severe sandstorms in north China. Starting in the late 1980s, the central government developed a national forest protection program (called the National Greening Campaign). When massive flooding of the Yangtze River occurred in 1998, this program was strengthened to include a complete logging ban in the upper reaches of the Yangtze and upper and middle reaches of the Yellow River.

Sand and desert (including areas classified as grasslands) extend over about 27 percent of China's land area, and this expanse has increased in the contemporary period. Desertification dries up rivers and lakes (leading to salinization of the soil, which cannot be used to grow crops), shriveled plants and vegetative cover, and reduced ground water levels (Li et al., 2004, 30-7). Much desertification can be attributed to deforestation as well as poor protection and overutilization of water resources in arid regions. Agricultural, commercial, industrial and residential developments also have an impact. A growing concern is grasslands degradation, where degradation of grass and soils despoils entire ecosystems (Han, 2001, 2-7).

The state's response has been comprehensive, including a total of seven different afforestation programs since the 1970s. These include establishment of plantations in the north, northeast and northwest China, protective afforestation in the upper and middle reaches of the Yangtze, development of a coastal shelterbelt, cropland protection and agro-forestry in the plains, afforestation of the Taihang Mountain as well as combating desertification (Yin et al., 2005, 19). A 2007 report announced that 20 percent of China would be forested by 2020 (Sun, 2007, 4). Forests play a critical role as a source of food in addition to staples for China's rural population. While foreign observers have been critical of the approach taken by the government, which among other problems increases monocultural plantations thereby limiting species diversification (Harkness, 1998, 924), the program overall has achieved some success in controlling flooding.

A related program of note is the sloping lands conversion program (SLCP; in Chinese *Tuigeng huanlin [cao]*), launched in 1999-2000 with a substantial government investment of about $50 billion. It was initiated after the 1998 Yangtze flooding to deal directly with erosion of lands in river basins that originally were forested, but had been converted to farmlands. With slopes greater than 25°, they were particularly subject to erosion, and government planners intended to convert 5.3 (later expanded to 14.7) million hectares of croplands on steep slopes to forest or grass coverage (Xu et al., 2004, 317).

Farmers received grain subsidies and cash payments for participation in the program, and for these reasons it became quite popular, but at a high annual cost. When grain prices spiked in late 2003 and 2004 in response to falling production, the Ministry of Land and Resources and several researchers hypothesized that SLCP led to the price rise (Ministry of Land and Resources, 2004). Most analyses, however, found that only 1 percent of the price increase in cereals could be attributed to SLCP, and that it had a large impact on reduction of the buildup of silt in irrigation networks and reservoirs and reduction in down-stream flooding (Xu et al., 2006, 130-48). In short, there did not appear to be a trade-off between land conversion and agricultural productivity. Further cost-benefit studies found evidence that the social welfare of farmers benefited from the program, but that the program was not cost-effective in achieving environmental goals—largely because of poor targeting of resources to areas of greatest ecological risk (Bennett et al., 2008, 196-220). The program continues but at a reduced expanse and cost.

AGRICULTURAL BIOTECHNOLOGY

The fourth response of the central government to concerns regarding agricultural production has incorporated the "scientific revolution" in agriculture: the combination of biological and technological techniques and methods for the purpose of improving seed varieties. The products of the biotechnological approach are called genetically-modified organisms or GMOs. They also are referred to as transgenic (*zhuanjiyin*).

China entered the era of biotechnology later than most nations, and its program began only in the mid-1980s. The primary reason for aggressive pursuit of agricultural biotechnology was to insure that food production kept pace with population growth (Zhang, 2000, 45-50). Other motivations included lessening damage of pesticides to the environment, participating in leading-edge technology (as a means of raising China's status scientifically, e.g., through the sequencing of the rice genome), and improving the competitiveness of China's many small farms.

Most of the Chinese research in agricultural biotechnology has been done in the public sector. The central government established National Key Laboratories, which were established to conduct research in the areas of agricultural biotechnology, plant breeding, and crop genetics. Laboratories were constructed in north, central and south China. The Ministry of Agriculture, the Ministry of Education, and the Chinese Academy of Sciences supported laboratories at many universities throughout the country (Keeley, 2006, 394). The Ministry of Science & Technology (MOST) has been most actively involved in the funding of research institutes. The investment grew quickly to exceed that of any other developing country. By 2003, China was spending about $350 million in plant biotechnology and $570 million in total agricultural biotechnology (plant, animal and microorganism) in terms of purchasing power parity, which was close to the amount that the US government spent (but overall American spending was higher because of significant private sector investment) (Huang et al., 2002, 674-77 and Rozelle et al., 2005, 93). Private sector firms played a small part in this enterprise with the exception of a few multinational agribusinesses such as Monsanto.

Cotton was the first transgenic crop to be commercialized in China, in 1997. China is the world's largest producer of cotton, but the plant is more affected by insect pests, specifically the bollworm, than any other crop grown in China. Scientists developed transgenic cotton varieties by inserting a toxin-producing gene from *Bacillus thuringiensis* (Bt), which made the plant resistant to bollworm infestation. In 2010, nearly 70 percent of China's cotton is grown from Bt cultivars. Bt cotton has increased cotton yields while greatly reducing pesticide spraying (as well as associated labor efforts). China has been the world's largest user of chemical pesticides which have huge costs in terms of human health and environmental degradation, and thus a two-thirds reduction in pesticide use in cotton-growing areas is a significant accomplishment. However, continued use of Bt cotton has been accompanied by a rise in incidence of secondary pests, which has dampened the ardor for this transgenic crop (Huang et al., 2004, 27-54 and Huang et al., 2003, 55-67).

In addition to Bt cotton, the central government also approved transgenic varieties of tomatoes, bell peppers, petunias, and insect resistant poplars and papayas. Such approvals have not had a significant impact on agricultural productivity. In fact, most attention has focused on cereal crops, with rice as the prime object of attention. Rice remains China's major food crop, and because of development of regulatory concerns (Pray et al., 2006, 148), a long period of field and environmental testing transpired before new Bt rice varieties with insect-resistance properties were brought to the pre-commercial clearance stage. Effectively, from 2002 to 2009, Bt rice remained in this ambiguous stage because of growing resistance globally and locally to GM products.

Areas of resistance included concerns about the potential impact of transgenic products on human health, on the environment, and on biodiversity, especially the risk of contaminating wild rice species. These concerns are found in other nations but have been most pronounced in Europe and with the European Union (EU). China's leaders and government agencies were divided too as to whether GM crops such as rice (and then maize and wheat) should be commercialized. Meanwhile the public is not well-informed on GM issues (Huang et al., 2006, 147-48; Xue, 2008, 101-09).

Then, in late 2009, the Ministry of Agriculture issued biosafety certificates for two strains of pest-resistant GM rice and one strain of maize. Huang Dafang, a member of the ministry's biosafety committee, commented that the strains would need to be registered and undergo production trials before commercialization could begin, and that this process would require from three to five years (Shan & Jin, 2010, 1).

The official justification for this breakthrough in the commercialization process was the need to increase grain production by 10 percent between 2009 and 2020 (currently China produces slightly more than 500 million metric tons of grain annually), so that food production would exceed anticipated population increases. Second, approximately 10 percent of the output from non-genetically modified rice is lost to diseases and pests, and use of GM technology could reduce this loss. Furthermore, the two rice strains, developed at Huazhong Agricultural University, would help reduce the use of pesticides by 80 percent and increase yields by as much as 8 percent (Wang & Shan, 2010, 2), and that this would increase farmers' income by $4 billion.

Response to this new development was mixed. Officials stated they would be "open-minded" about continued testing and production trials, but several scientists were critical. Zheng Fengtian, vice-dean of the Rural and Agricultural Development Institute at Renmin University said "If the world has not reached a consensus on GM food safety, it would be too proactive and risky to commercialize GM crops on a large scale"(Xinhua, 2010, 3). Yuan Longping, the developer of hybrid rice, warned that health implications of some GM crops and particularly the anti-pest strains were not clear; he said that such crops "need to undergo human trials for at least one or two generations"(Shan & Wu, 2010, 4).

Fang Lifeng, spokesman for Greenpeace China's GM program reported that its 2007 survey found that 65 percent of 2,000 people polled said they would not choose GM food because of safety concerns (Shan and Jin, 2010, 1). Shortly after the government announcement, Greenpeace released results of a survey it had conducted in nine supermarkets of eight cities, and claimed that transgenic rice was being marketed in Hunan Province (Shan, 2010, 4). (This environmental NGO had reported finding unapproved GM rice being illegally grown in Hubei province in 2005 [McBeath & McBeath, 2010, 190].) Government officials were skeptical about this account. It seems likely that the issue of Bt rice will stimulate both public and elite discussion and debate before the strains are registered and finally approved for commercialization.

NATIVE GERM PLASM IMPROVEMENTS

The fifth government program we consider concerns native species of plants whose growth is enhanced through plant breeding. For hundreds of years, Chinese farmers as well as agriculturalists developed plant breeding programs based on the selection of plant lines increasing crop production; through happenstance, some of these had high yield and others had disease- and drought-resistance properties. They carefully selected seeds from crops that produced the most desirable characteristics; and they produced hybrid species through grafting superior elements of one variety onto another.

Breakthroughs in the science of genetics in the twentieth century made the breeding of highly productive and resistant varieties systematic, and agricultural stations disseminated the seeds of these varieties. In the reform period, sales from seeds and chemical pesticides became major income sources supporting operations of many provincial or country agricultural stations. Because of keen competition and incomplete intellectual property protection, new varieties of crops are released almost every year, and some of them are not well tested. Recently, awareness of damage to the environment through intensive use of chemical controls also has increased interest and support of programs breeding plants for disease resistance.

Given a choice, breeders and farmers usually select varieties possessing traits such as high yield, early maturing and semi-dwarf and not disease resistance, because diseases are perceived to be manageable by chemical pesticides and other means. The greatest government investments have been for the purpose of improving agricultural productivity. For example, the Ministry of Agriculture has invested heavily in "super rice," and one hybrid developed in the early 1970s by Yuan Longping, a scientist in the Chinese Academy of Sciences, has been successful in increasing rice productivity. Supporters of his work claim that the hybrid he developed "boosted the country's rice output over the ensuing 25 years by a combined 400 million tons" (Wang, 2007, 3). In 2008, one of Yuan's colleagues, Qing Xianguo, said the innovation in hybrid rice technology would increase production 20 percent annually (Wu, 2008, 3).

However, productivity of other hybrids has not yet met expectations for greatly improving grain yields, as revealed in the decade during which the Ministry of Agriculture has conducted research and development on highly productive rice species. Most of the three dozen varieties of super rice are strains for just a single-season harvest. They thrive only in high-yield farmlands where natural conditions are optimal, and not under normal conditions. Further, inadequate auxiliary planting and management expertise limit productive capacity (Zhao, 2006, 7).

A related factor is demand for improvement in quality of food products and in particular the quality of cereal crops. The release of new high yield cultivars and hybrids has not been accompanied by much improvement in the eating and cooking quality of the products. In fact, one critic comments: "Most of the widely used cultivars and hybrids have poor cooking and eating qualities, and thus are disfavored by producers and consumers" (Zhang, 2000, 45).

As we note below, agricultural research in China has great opportunities to improve native germ plasm for disease resistance.

PLANT DISEASE AND PEST CONTROL

In this sixth area, we consider the impact of plant diseases and insect pests on agricultural production in China, and the prevalent means used to control their effects on cereal crops. The dimensions of disease problems are reflected in recent government reports. The *2006 China Agriculture Yearbook*, prepared by the Ministry of Agriculture, reported that damages caused by economic insects, and escalation of disease and insect pest infestation had reached 342.8 million hectares (which repeats the count of lands double- or triple-cropped), an increase of 16.5 million hectares over the previous year. Losses to grains were 59.1 million tons. With 484 million metric tons of grains harvested in 2005, the loss was about 12 percent of total grain production. The major diseases listed were rice blast, wheat rust, and soybean stem rot. The most important insect pests were the brown rice planthopper, rice borer, cotton boll weevil, corn borer, and migratory locust (Ministry of Agriculture, 2006, 376).

In December 2006, the China Agricultural Extension and Service Center revealed that in recent years, despite redoubled efforts by the government to improve disease and pest monitoring and control measures, damages caused to agricultural crops by diseases and pests still reached 25 million metric tons per year. The increase in scope and severity of diseases and insect pests caused economic losses of $8.4 billion annually (Wu, 2006). This report gave several examples of the seriousness of disease and pest outbreaks:

- Eight consecutive years of migratory locust epidemics in agricultural regions
- Six consecutive years of stem borer outbreaks
- Six consecutive years of rice brown planthopper epidemics at detrimental scales in rice growing regions of the south
- Six consecutive years of wheat stripe rust epidemics
- Three consecutive years of rice blast outbreaks (Wu, 2006).

These disease and pest infestations present daunting challenges to the state, and China has a plentiful array of mitigation measures, which have reduced the impact on food production. We discuss chemical and biological control measures and then say a few words about breeding for disease resistance.

In China, chemical means are preferred by farmers because they produce quick results. Yet increasingly, highly toxic, systemic pesticides are applied, which are harmful to human and animal health. Farmers use chemical pesticides to reduce pathogens before the inoculum reaches the plant, for example by fumigating soils. Farmers also spray them on plant surfaces to protect them from infection. Fungicide seed treatment and foliar spray on stems are the most common practices for addressing plant disease, however many pathogens such as rice blast have developed resistance to chemical treatments after continuous use. Similar observations apply to resistance of insect pests such as rice planthoppers and aphids to frequent applications of insecticides. China has the world's highest application rate per hectare of chemical pesticides.

One large obstacle to improvement of disease management at the local level is the inherent conflict between objectives of China's agricultural stations. These are outreach program offices with few books, little equipment, and no tools to conduct fundamental scientific work. Instead, they have become enterprises relying for revenue on the products they sell farmers. While they sell seeds, including those with some disease-resistance properties, they make more money by selling pesticides, which farmers purchase because they produce fast results.

Biological controls involve using one biological species against another, and they have been part of the Chinese agricultural system since time immemorial. For instance, using ducks to control weeds in rice paddy fields and using ducks and chickens to eat grasshoppers and locusts are still part of the pest management regime. The development of biological controls today involves using antagonistic microorganisms and predators to mitigate plant diseases and insect pests by encouraging competition for space and nutrients and by inhibiting, parasitizing and potentially eradicating pathogens and insect pests.

Many biological control agents have been developed and commercialized in China. Some are sold as biopesticides, others as soil amendments and growth enhancers. In addition, natural enemies of insect pests such as aphids, Asian corn borer and locusts have been identified and cultivated, including egg, larval and pupal parasitoids (McBeath & McBeath, 2010, 153). The advantage in use of biological controls is that they cause little or no damage to the natural environment or to human health. The disadvantage is that they are slow-acting. Similar to chemical pesticides, the efficacy of biological pesticides is varied, but because of environmental concerns, they increasingly are being accepted.

Finally, as indicated in the previous section, plant varieties are now often bred for resistance against important pathogens, such as rice blast, wheat rust, wheat powdery mildew and corn leaf blight. Researchers have made rapid advances in analysis of resistance reaction of rice germ plasms to rice blast as well as many other pathogens (Yuan et al., 2005, 27-34). However, diseases such as rice blast possess great genetic variation, and new virulent strains/races cause a breakdown of resistance in crops, necessitating development of new resistant plant varieties.

Advances in plant disease and insect pest research have barely kept pace with infestations, and the loss of agricultural production to this cause remains serious.

Farm Cultural Practices

No nation can claim greater historic attention paid to improvement of farming practices than China. In the early Han Dynasty, officials emphasized development of a class of independent small farmers, agricultural extension and compilation of materials on agricultural improvement, support of irrigation systems and government granaries (Anderson, 1988, 44). Historian Mark Elvin noted that in the Song Dynasty, there were three important agricultural innovations:

[1] Farmers learned to prepare their soil more effectively as the result of new knowledge, improved or new tools, and the more extensive use of manure, river mud and lime fertilizers.

[2] Strains of seed were introduced which either gave heavier yields, or resisted drought better, or else by ripening more rapidly made it possible to grow two crops a year on the same land.

[3] A new level of proficiency was reached in hydraulic techniques, and irrigation networks of unprecedented intricacy constructed (Elvin, 1973, 118)

Few other changes were made until the end of the dynastic era in the early twentieth century. In the Republican era, agricultural practices modernized. Historically, Chinese agriculture has been highly labor-intensive, but by the 1930s, machines with diesel engines were introduced in several southern regions. Traditional Chinese agriculture employed natural means to fertilize crops, particularly use of human and animal manure. In the early Republican era, chemical fertilizers were introduced to many Chinese farms, and by the 1940s, fertilizer plants were springing up along the eastern coast. Both mechanization of agriculture and chemical fertilizers and pesticides improved agricultural productivity. Also, improved crop seeds were developed with disease- and drought-resistant properties (Buck, 1973, 66-67).

In the Maoist era of the early People's Republic, officials made radical changes in agricultural organization, which had devastating effects on agricultural production. However, the technological revolution restarted, with further development of mechanization and improvements in irrigation, rural electrification, fertilization, and crop breeding. The regime began the dissemination of new farming techniques, and by the early 1960s, more than 100,000 staff worked out of extension service stations (Chao, 1970, 89).

After economic reforms began in the late 1970s, farm families had new incentives to increase production and lower costs. Increased agricultural research and the improvement of agricultural infrastructure led to increases in productivity as well. Yet the small size of most farms inhibited mechanization improvements (there are about 220 million farms in 2010). Also, demographic changes presented problems. Already less well-educated than urban Chinese, the better educated farmers were more likely to join the out-migration from rural areas, leaving a disproportionate number of very young and very old farmers in rural areas, and women were more numerous than men. In fact, at least 70 percent of the farmers joining the migrant labor pool have been men (Xinhua, 2008, 3).

Much more technical assistance is available to farmers today to disseminate new techniques and aid in improvement of farming practices (Li, 2007). The Chinese extension system has five levels, from the national through provincial, municipal, and county to the township. At this lowest level, about 400,000 staff work from 50,000 extension units, and they have most frequent encounters with farmers, spreading information on improved methods of crop cultivation, including fertilization and new seed varieties. About one-fifth of villages have farmers' service organizations, and there are, nation-wide, about one million farmer technicians. But most villages lack farmers' organizations and easy access to extension assistance.

While dissemination of information on new agricultural technologies has spread broadly, farmers in most areas have not yet changed their practices to reduce environmental degradation and threats to human health (especially from improper pesticide and fertilizer use). For example, a recent survey of farm households in the North China Plain found that only about 6 percent followed recommendations of extension agents regarding balanced input use (Zhen, 2005, 507-22).

Cultural control measures today include a broad and diverse collection of production practices such as crop rotations, soil tillage, changing the time of planting and plant spacing, providing soil drainage and making other irrigation system changes and improving sanitation. Significant advances have been made in controlling plant diseases and insect pests by mixture planting and inter-planting of crops (Shen et al., 2007). Also, increasingly, scientists have paid attention to ecological regulation strategies against crop pests, and to biodiversity-rich cropping practices (Zhu; Guo, 2006, 1-4). Finally, polyethylene mulching is used extensively in China for the purposes of raising soil temperature, conserving water, and eradicating many soilborne pathogens. Tunnels of various heights are used to protect plants from infection by airborne pathogens, and also to conserve heat and water and extend the season.

These modernizations have had a positive impact on agricultural production.

CONCLUSION

Food sufficiency is a priority of the Chinese state, and leaders watch production figures very carefully. China's ability to provide most of its cereals' needs (it is about 95 percent self-sufficient in staples) is an indication of the success of the reform leadership.

In the early twenty-first century, reports indicated that state spending on agriculture, in support of agricultural production, science and technology, capital construction, and poverty relief totaled $23.4 billion—about 7 percent of government spending then (as compared to 10 percent in 1990). The state spent an additional $827 million on price subsidies for grain, cotton, and edible oil (Waldron et al., 2006, 287). Spending on agriculture exceeded that on national defense (as defined by government statistical reports), but lagged behind expenditures in the areas of education, health, and capital construction. However, the real growth rate of spending on agricultural research has slowed in China, a pattern noticeable globally (Parday & Beintema, 2001). In 2003, China's investment in agricultural R&D was 0.4 percent of agricultural output, compared to 3 percent in economically developed countries and 1 percent on average internationally (Ministry of Science & Technology, 2007). And the ratio of public to private sector spending is still far higher than in most other countries.

In this chapter we have examined seven responses of the state to perceived and actual threats to agricultural production: attempts to restrict the continued loss of arable land; investments in irrigation systems and re-distribution of water to deal with scarcity in the north and west; control of erosion, desertification, deforestation (and consequent flooding) through a massive re- and afforestation program; large-scale investment in agricultural biotechnology and building the capability to use it to improve productivity; improvements in native germ plasm; development of plant disease and insect pest controls; and enhancements in farm cultural practices.

In each area of inquiry, we noted costs and benefits to the approach taken by state agencies, by bureaucrats, planners, scientists, and farmers themselves. Policy choices of leaders reflect several stimuli: historical inertia in a large and old agricultural country, the core imperative of the state to demonstrate effectiveness by insuring self-sufficiency in cereals, the need to maintain legitimacy of the state system as it undergoes extensive marketizing reforms, and competitive pressures in the international agricultural trade system.

Our more critical observations (based on personal experience) center on the last two state responses of plant disease and insect pest control and farm cultural practices. The state has invested large resources in basic scientific research and development of institutes, and many researchers conduct cutting-edge research. However, little funding has been allocated to applied research and even the dissemination of this research to the field is problematical. For example, entomologists may have a poor understanding of disease transmission properties of insects. On the other hand, plant breeders are ignorant of most diseases of the crops they are breeding. This represents an absence of cross-pollination and interdisciplinarity.

We mentioned that local agriculture stations lack much information on plant disease and insect pest conditions; they have no laboratory facilities. Staff members of these stations, who have direct contact with farmers, are poorly supported and trained. It is not uncommon for them to regard symptoms displayed on diseased plants as "normal." Extension funding needs to be directed to the improvement of plant disease and inspect pest identification and to the development of means of disease and pest control that are effective, environmentally benign and safe for humans and animals.

REFERENCES

Anderson, E. N. (1988). The food of China. New Haven, CT: Yale University Press, 44.

Bennett, J., Wang, X. & Zhang, L. (eds) (2008). *Environmental protection in China: land-use management.* Cheltenham, UK: Edward Elgar, 192-220.

Brown, L. (1995). *Who will feed China? Wake-up call from a small planet.* Washington, DC: Worldwatch Institute.

Brown, L. (1998, July 1). *China's water shortage could shake world security.* World Watch.

Buck, J. L. (1973). Land and agricultural resources. In: Wu, Y. (ed). *China: a handbook..* New York: Praeger, 66-67.

Chao, K. (1970). *Agricultural production in communist China*: 1949-1965. Madison, WI: University of Wisconsin Press, 89.

Di, L. (2004). Land-use patterns and land use change. In: Hsieh, C. M. & Lu, M. (eds). *Changing China: a geographical appraisal.* Boulder, CO: Westview Press.

Economy, E. (2004*). The river runs black: the environmental challenge to China's future.* Ithaca, NY: Cornell University Press, 126.

Elvin, M. (1973). *The pattern of the Chinese past.* Stanford, CA: Stanford University Press, 118.

Fan, S., Zhang, L. & Zhang, X. (2004). Reforms, investment, and poverty in rural China. *Economic Development and Cultural Change*. Vol. 52, no. 2: 417.

Greenhalgh, S. & Winckler, E. A. (2005). *Governing China's population: from Leninist to neoliberal biopolitics.* Stanford, CA: Stanford University Press.

Guo, Y. (2006). Illustrations with real examples of using ecological regulation strategies against crop pests in China. *Plant Protection*, Vol. 32, no. 2: 1-4.

Han, N. Y., Jiang, G. & Li, W. J. (2002). *Management of the degraded ecosystems in Xilingol biosphere reserve.* Beijing: Tsinghua University Press, 2-7.

Harkness, J. (1998). Recent trends in forestry and conservation of biodiversity in China. *China Quarterly* 156: 914.

Ho, S. P. S. & Lin, G, C. S. (2004*).* Non-agricultural land use in post-reform China. *China Quarterly* 179: 778.

Huang, J. K., Pray, C., Rozelle, S. & Wang, Q. (2002). Plant biotechnology in China. *Science* 295 (5555): 674-677.

Huang, J. K., Hu, R., Pray, C., Qiao, F. & Rozelle, S. (2003). Biotechnology as an alternative to chemical pesticides: a case study of Bt cotton in China. *Agricultural Economics* 29: 55-67.

Huang, J. K., Hu, R., van Meijl, H. & van Tongeren, F. (2004). Biotechnology boosts to crop productivity in China: trade and welfare implications. *Journal of Developmental Economics* 75: 27-54.

Huang, J., Hu, R., Rozelle, S. & Pray, C. (2005) Insect-resistant GM rice in farmers' fields: assessing productivity and health effects in China. *Science*, Vol. 308 (April 29): 688-90.

Huang, J., Qiu, H., B. J. & Pray, C. (2006). Awareness, acceptance of and willingness to buy genetically modified foods in urban China. *Appetite*, Vol. 46: 147-48.

Huang, Q., Rozelle, S., Lohman, B., Huang, J. & Wang, J. (2006). Irrigation, agricultural performance and poverty reduction in China. *Food Policy* 31: 30-52.

Keeley, J. (2006). Balancing technological innovation and environmental regulation: an analysis of Chinese agricultural biotechnology governance. *Environmental Politics* 15 (2): 394.

Li, C. (2007). *Reform of China's agricultural technology structure*. Unpublished paper. Beijing.

Li, F. (2007, April 23). Official: more than 10% of arable land polluted. China Daily, 1.

Li, F. (2007, June 26). Minimum land bank to stay in effect. *China Daily*, 3.

Li, X. Y., Yang, J. & Wang, L. X. (2004). Quantitative analysis on the driving role of human activity on land desertification in an arid area: the case of the Tarim River Basin (in Chinese). *Resources Science*, Vol. 26, no. 5 (September): 30-7.

Ma, J. (2004*). China's water crisis.* Norwalk, CT: Eastbridge, ix.

McBeath, J. & McBeath, J. H. (2009). Environmental stressors and food security in China. *Journal of Chinese Political Science* 14: 49-80.

McBeath, J. H. & McBeath, J. (2010). *Environmental change and food security in China.* Dordrecht, NE: Springer.

Moody, A. & Lan, L. (2010) Focusing on future urbanization. *China Daily*, March 22, 8.

Parday, P. G. & Beintema, N. M. (2001). Slow magic: agricultural R&D a century after Mendel. *IFPRI Food Policy Report*. Washington, DC.

People's Republic of China (PRC), Ministry of Land and Resources (2004). 2003 *China national report on land and resources, Beijing.*

PRC, Ministry of Agriculture (2006). 2006 *China agriculture yearbook*, 376.

PRC, Ministry of Science & Technology (2007). *2006 report on technological development of China's villages (in Chinese*). Beijing: China Agricultural Publishers.

PRC, State Council (2004). *China South-to-North Water Diversion Project*. Beijing, Construction Committee of the South-North Water Diversion Project, 15.

PRC, (2007*). Statistics year book of China*. Beijing: China Statistics Press.

Pray, C. E., Ramaswami, B., Huang, J., Hu, R., Bengali, P. & Zhang, H. (2006). Costs and enforcement of biosafety regulations in India and China. *International Journal of Technology and Globalization* 2(1/2): 148.

Qian, Y. (2010). Water project alters flow of life along its route. *China Daily,* March 4, 1.

Ross, L. (1988*) Environmental policy in China*. Bloomington, IN: Indiana University Press.

Rozelle, S., Huang, J. & Otsuka, K. (2005). The engines of a viable agriculture: advances in biotechnology, market accessibility and land rentals in rural China. *China Journal* 53: 93.

Shan, J. (2010). Greenpeace survey says modified rice on sale. *China Daily*, March 16, 4.

Shan, J. & Jin, Z. (2010). GM rice could hit shelves. *China Daily*, February 4, 1.

Shan, J. & Wu, J. (2010). Modified grain not set for market. *China Daily,* March 11, 4.

Shapiro, J. (2001*). Mao's war against nature: politics and the environment in revolutionary China*. Cambridge, UK: Cambridge University Press.

Shen, J. H., Nie, Q, Huang, D. R., Liu G. J. & Tao, L. X. (2007). Recent advances in controlling plant diseases and insect pests by mixture planting and inter-planting of crops (in Chinese). *Acta Phytopathologica Sinica*, Vol. 34, no. 2.

Smil, V. (1993) *China's environmental crisis: an inquiry into the limits of national development.* Armonk, NY: M.E. Sharpe.

Smil, V. (1999). Research note: China's arable land. *China Quarterly* 158: 423-4.

Sun, X. (2007). More nature reserves to protect forests. *China Daily*, December 5, 4.

Turner, J. & Otsuka, K. (2006*). Reaching across the water.* Woodrow Wilson International Center for Scholars, Washington, DC., 1.

Waldron, S., Brown, C. & Longworth, J. (2006). State sector reform and agriculture in China. *China Quarterly* 186: 282.

Wang, Q. (2008). Stimulus sops up farmland. *China Daily*, December 31, 3.

Wang, Y. (2007). Secrets of hybrid crops in focus. *China Daily,* October 15, 3.

Wang, Z & Shan, J. (2010). GM push sparks debate: science to boost food supply. *China Daily*, February 4, 2.

Wong, E. (2008*). China may let peasants sell rights to farmland.* New York Times, October 11, A1.

Wu, J. (2006). China harmful biologicals cause loss of 57 billion yuan (in Chinese*). China Chemical Engineering News,* December 6.

Wu, J. (2008). Hybrid rice set to boost yield by 20%. *China Daily*, July 24, 3.

Xie, C. (2007). 1.7m hectares of additional arable land "by 2020." China Daily, June 22, 2.

Xinhua (2007). Salt lands may offer solution to land shortage problem. *China Daily*, December 27, 3.

Xinhua (2008, February 22). Farming losing its appeal. *China Daily,* 3.

Xinhua (2008, October 11). Move to minimize disputes over land use. *China Daily*, 1.

Xinhua (2008, October 20). Farmers get leeway on use of land. *China Daily*, 1.

Xinhua (2010). Modified food issue debated. *China Daily*, March 4, 3.

Xu, Z.,Bennett, M. T., Tao, Ran & Xu, J. (2004). China's sloping land conversion programme four years on: current situation and pending issues. *International Forestry Review*, Vol. 6, nos 3-4, 317.

Xu, Z., Xu, J., Deng, X., Huang, J., Uchida, E. & Roselle, S. (2006). Grain for green versus grain: conflict between food security and conservation set-aside in China. *World Development* 34(1): 130-48.

Xue, Dayuan (2008). Environmental impacts and safety regulation of genetically modified organisms (in Chinese). *Beijing: China Environmental Science Press*: 101-09.

Yardley, J. (2007). *Chinese dam projects criticized for their human costs*. New York Times, November 19.

Yin, R., Xu, J., Li, Z. & Liu, C. (2005). China's ecological rehabilitation: the unprecedented efforts and dramatic impacts of reforestation and slope protection in western China. Woodrow Wilson Center. *China Environmental Series* (7): 19.

Yu, X. (2008). Land erosion "threat to food supply." *China Daily*, November 23, 2.

Yuan, X. P., Wei, X. H., Yu, H. Y., Wang, Y. P. & Tang, S. X. (2005). Analysis of the resistance reaction of some Chinese rice germ plasm (*Oryza sativa L.*) to rice blast. *Plant Protection,* Vol. 31, no. 2: 27-34.

Zhang, Q. (2000). Agricultural biotechnology opportunities to meet the challenges of food production. In: Persley, G. J. & Lantin, M. M. (eds). Agricultural biotechnology and the poor: proceedings of an international conference. *Consultative Group on International Agricultural Research,* Washington, DC, 45-50.

Zhao, H. (2006). Higher yields sought from the super rice. *China Daily*, October 7-8, 7.

Zhao, H. (2007). No yielding on arable land: minister. *China Daily*, July 13, 1.

Zhen, L., Routray, J. K., Zoebisch, M., Chen, G., Xie, G. & Cheng, S. (2005). Three dimensions of sustainability of farming practices in the North China Plain. *Agricultural Ecosystems and the Environment* 105(3): 507-22.

Zhu, Y. (no date). *Cultivating biodiversity for disease control, a case study in China*. Yunnan Agricultural University.

In: Agricultural Production
Editors: Felix C. Wager, pp. 181-199

ISBN: 978-1-61668-695-6

Chapter 9

SOCIAL LABELS FOR AGRICULTURAL PRODUCTS – THE EXTRINSIC QUALITY DIMENSION OF FAIR TRADE CERTIFICATES

Sebastian Jaenichen and Torsten Steinrücken
Technical University of Ilmenau (Germany)
Thuringian Ministry of Economics

The concept of Fair Trade is applied to the marketing of a variety of goods. In recent years it has met with a continually increasing interest of demanders. Different Fair Trade organisations are trying to accomplish an improvement of working and living conditions in countries where agricultural products stem from. They use Fair Trade certificates and pay a price evidently above world market standard. This trade concept is meant to permit basic social standards to be observed, especially in agricultural production. Fair Trade organisations make use of extrinsic product quality. Extrinsic product quality determines among other factors the consumer's perception of product quality. An important though thitherto poorly considered factor of extrinsic product quality is when the consumer cares about the production process, even if it does not affect product quality. The article deals with the questions how Fair Trade works and if the social aims in focus can be achieved by this trade concept. Our main result is that even though efficiency of redistribution through the Fair Trade institutions is lower than through traditional relief organisations the Fair Trade concept provides an additional incentive to pay for better living conditions in the third world. Fair Trade initiatives are unlikely to crowd out other redistribution channels. Moreover, there is a stimulus for producers to reorganize the production process in a more social manner even if this is not awarded by the fair trade company.

1. PRELIMINARY REMARKS

The increasing globalisation of world trade and its consequences have been criticized worldwide. Most evident token of this development are the annual protest actions on the occasion of the leading economies' conferences like the World Economic Forum in Davos. Another sign is the foundation of organisations like ATTAC who are calling for a change in international trade. Yet there have been recent initiatives which try to change the trade with certain goods. The activities of different Fair Trade[1] organisations like TransFair or Max-Havelaar try to accomplish an improvement of working and living conditions for producers in developing countries by means of prices distinctly above the world market level. The efforts of these organisations are supported by the demanders and their consumer behaviour. It is true that at the moment the scope of fairly-traded products is still relatively narrow. But taking into account the growth rates of this product segment and the tendency towards an increasing interest on the demanders' side, one has to concede that the analysis of this trade concept gains in significance.[2] This paper will investigate the Fair Trade initiatives from an economic point of view and will answer the question if the social goals envisaged by these initiatives can be achieved by this trade concept. To begin with, the development of the Fair Trade concept will be illustrated. In the following, the stimulating effects on demanders in developed countries and producers in developing countries will be investigated thoroughly.

2. Distribution of Agricultural Products Through the Concept of Fair Trade

2.1. The Development of Fair Trade

The basic idea of Fair Trade goes back to the 1950s. This alternative trade concept has developed from the Western developing countries' increasing awareness of the fact that the different countries and social classes do not necessarily benefit to the same extent from growth of trade and economy. In many European countries, alternative trade organisations developed which tried to counteract the structural disproportions in trade between developed and developing countries.[3] Creating alternative distribution channels was an attempt to set a different trade concept against the structure of world trade which was regarded as unjust. The idea of Fair Trade can be understood as a form of ethic action since it requires the companies to observe their ethic and social responsibility concerning trade with developing countries. Fair Trade is designed to help to abolish disparities between developed and developing countries which have occurred due to the relative fall in raw material prices, particularly concerning agricultural products. The professed aim is to ensure the producers of a share of the total profit which matches the appropriate proceeds for their preliminary work. As a rule, this is done by the Fair Trade partners' agreeing upon a fair price.[4]

1 Fair Trade initiatives especially work with coffee. Coffee was the first fairly-traded product. In recent years further agricultural products like tea, cocoa, sugar, honey, chocolate, bananas, orange juice, olive oil, rice and even industrial products such as sport balls have been following.

2 In the last years the international umbrella organisation, the Fair Trade Labelling Organisation (FLO) could register 20 per cent growth. In 2002, 600 traders in 17 consumer countries have earned about 300 million Euros with Fair Trade products. Cf. www. transfair.org.

3 In Germany the „Aktion Dritte-Welt-Handel" was founded. It especially criticizes the West's affluent society and its continuing growth. For a detailed account of the development of the Aktion Dritte-Welt-Handel and this organisation's goals cp. Ottowitz (1997), p.5.

4 Cf. Kommission der Europäischen Gemeinschaften (1999), S. 4f..

2.2. Distribution Channels of Fair Trade Products

Apart from the sale of fairly-traded products via world shops which has already been established for a long time, the sale of products with a Fair Trade certificate has developed greatly in recent years. The products are labelled with special social labels and certificates. They are distributed via traditional distribution and marketing channels or via the internet. By using the conventional marketing and retailing structures producers in developing countries gain access to the developed countries' market. Fair Trade products become available to a wider range of consumers. In case of sale via third world shops the identity of the sales organisation serves as a guarantee for the products and the business practices meeting the Fair Trade principles. But if fairly-traded products are certificated and sold via conventional trade organisation, no such distinct mutual trust between consumers and the selling organisation can develop. Only the Fair Trade Label indicates that production and sale of the respective product have been following the guidelines of Fair Trade. By putting their label on a product, the different Fair Trade organisations[5] guarantee that the production in the country of origin has taken place under the observance of the respective social standards. The organisations do not trade but only give out the social labels and control the observance of social standards.

2.3. Goals of Fair Trade Initiatives

The Fair Trade concept is meant to allow producers in developing countries to observe basic social standards in production. Most important element of the Fair Trade idea is to raise the producers' wages for their products to a price which to a certain extent clearly resides above world market price.[6] In addition to this guaranteed price, the producers receive an extra price bonus by some Fair Trade organisations if they produce in an acknowledged ecological way.[7] Apart from the guaranteed sales price further benefits for the producers consist in long-term supply relationships and the right to be pre-financed. By means of paying a price distinctly above world market standard it is tried to provide an incentive for producers to participate in Fair Trade and by these means to achieve the diverse social aims and the aims of development politics. These aims are, e.g.: (1) to promote especially disadvantaged peasant families and support self reliance initiatives as well as concrete improvements for plantation workers and harvesters (2) to prohibit forced labour and child labour, (3) to obey internationally obtained employment protection guidelines, (4) to support a long-term development process of great sustainability, and (5) to provide help for conversion to non-

---

5 There are various initiatives and social labels which indicate the observance of particular social standards during production in developing countries. Examples are the labels Rugmark (carpets), Fian (cut flowers) or Palmpool (washing agents). Cf. Weißmann and Wirtz (2001), p. 19.

6 At the moment the minimum price for conventional Arabica coffee which is paid by Fairtrade amounts to \$1,26/lb (for organic coffee 1,41/lb), whereas the world market price in January 2004 was \$0,73. (If the world market price rises above the TransFair minimum price, an advance of at least 5 per cent above the world market price is paid.) The world market price for coffee is subject to great fluctuations: In 1994, the New York produce exchange recorded a price of 202 US cents/lb (1 lb = 0, 4536 kg), in 2002 it was only about 45 US cents per lb. For the current coffee price cf. International Coffee Organization (2004).

7 Cf. Weißmann and Wirtz (2001), p. 19.

polluting methods of cultivation and processing.[8] It has to be admitted that the producers only profit from the advantages of Fair Trade conditions if they are a member of one of the Fair Trade organisations. Fair Trade initiatives do not make use of minimum social standards, which many developing countries reject to introduce.[9] Minimum social standards tend to impose a ban on imports and sale on those producers who do not keep to the requirements. Instead, Fair Trade initiatives try to provide a positive incentive. As a tribute to the obedience of particular standards of production, Fair Trade organisations reward the producers with a payment above the market price. This price premium is meant to ease the conversion to more social methods of production[10].

3. Economic Considerations about the Fair Trade Idea

3.1. Demanders' Interest and Relative Prices

Most economic analyses act on the assumption that human being is a *homo oeconomicus*[11] and that an individual orients towards expected subjective utility in his behaviour. The model for maximisation of expected subjective utility under constraints declares that an individual will pick from a given set of alternative activities the one which promises the greatest utility. At first sight it seems surprising that in case of relatively homogeneous goods like coffee or bananas some consumers decide in favour of the more expensive product. Traditionally distributed products and Fair Trade products hardly differ with respect to their functional utility. Regarding material, certificated and non-certificated products show at most very little divergence, they merely differ concerning the process of production.[12] However, the perceived product quality does not depend on these objective product characteristics, which are commonly named *intrinsic product quality factors,* alone. Even if we disregard intrinsic product quality, consumers can still have a preference for Fair Trade products due to what Antle (1999) counts among extrinsic quality. This is when the consumer cares about the production process, even if it does not affect product quality. Sometimes products have their origin in the same plantations and have only been distributed

8 See, for a detailed account of the Fair Trade criteria Kunz (1999), Welford, Meaton and Young (2003). Further information on www.transfair.org.

9 For aspects of social minimum standards concerning trade see Busse und Grossmann (2003).

10 Not only the Fair-Trade-Organizations but also some business companies are prepared to pay a price above the world market level (One example is the franchise-company Starbucks, that pays for coffee more than the world market price). Such behaviour could be explained on the one hand with efficiency arguments and on the other hand with the same arguments we presented above. Like a consumer, a business company can also pay for social production methods. This could generate a pay off because the company gets a positive (social) image and consumers may be willing to pay for the use of fair traded products by the company.

11 Economic analyses are not concerned with real individuals who show many different behaviour patterns. In fact, they are based on the hypothetic construction of homo oeconomicus, which is characterised by a limited number of behaviour patterns and character traits which are postulated a priori. Cf. Mummert (1995), p. 38.

12 Because producers often cannot market their entire produce according to Fair Trade conditions, the remaining amount is sold via traditional distribution channels at world market price. Cf. Liebrich (2003), p. 35.

in a different way.[13] Thus, perceived product quality is a composition of intrinsic (rather objective) and extrinsic (rather subjective) product quality factors.

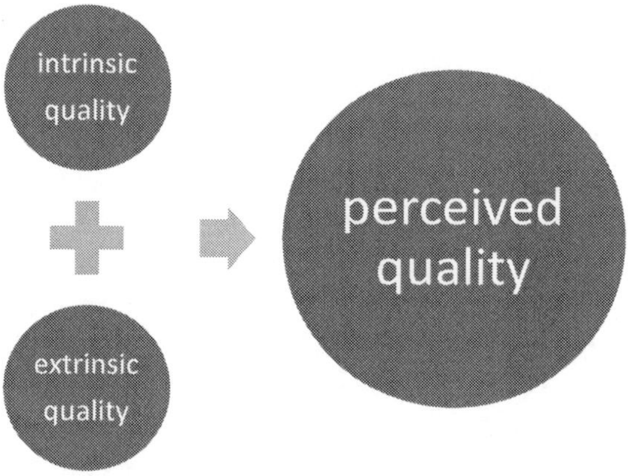

Dimensions of perceived product quality

The above figure assumes a dichotomy of quality dimensions which can be explained by the consumer's difficulty to judge product quality on the basis of intrinsic cues alone which are summed up in the right column of the following table. However, the increasing popularity of Fair Trade certificates cannot be plausibly attributed to a need of reduced information requirements on the consumer side of the market. It rather traces back to an expanded interest in the social environment of agricultural production and thus to an interest which indicates the consumers ability to handle further information of that type.

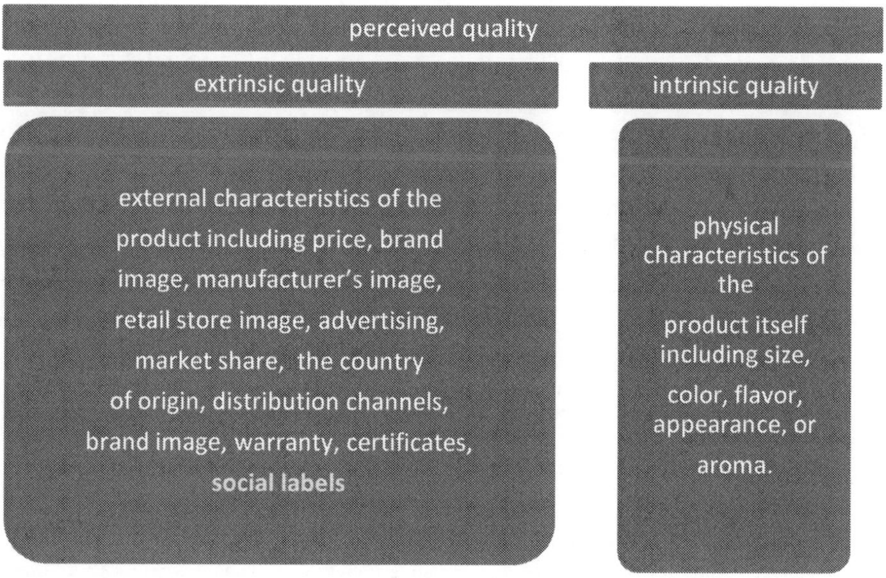

Dimensions and indicators of perceived product quality

13 Piepel (2000), p. 8.

The observable consumer behaviour can partly be explained with the help of economic theory. Rational consumers will opt for buying a relatively expensive Fair Trade product if its net utility is higher than that of a conventionally distributed product. In order to explain this behaviour it seems sensible to distinguish between a direct functional utility consisting in consumption of the respective goods and an additional utility. Both quantities can be recorded in the *homo oeconomicus'* utility function. Let us assume that the direct gross functional utility from the consumption of a product is the same with fairly-traded and conventionally distributed goods. Consuming a Fair Trade product, the consumer does not only profit from the consumption. Extra components like supporting the producer or organic farming provide a further utility to the consumer. In a study carried out by OTTOWITZ (1997) 74.8 per cent of the purchasers of fairly-traded coffee declared that their purchase pattern was determined by the wish to support the producers. 15.4 per cent named organic farming. Only for 6.5 per cent of the purchasers taste is the decisive factor for buying fairly-traded coffee. If the additional utility, which is only connected to the consumption of fairly-traded products, exceeds the utility losses caused by the supplementary charge for the Fair Trade products, a rational consumer will demand such a product. By linking the consumption of a product with benefits for social or ecological activities, one can cause a higher willingness to pay in certain groups of demanders.[14]

From the economic point of view, Fair Trade products are a product bundle. The acquisition of the conventional product is linked with a donation for social or ecological methods of production. Since both components can be obtained separately, this is called mixed bundling.[15] Consumers could also buy conventionally distributed products and express their social or ecological dedication by a donation to the respective relief organisation. In comparison with the separate demand for each of the product components, the joint purchase of both components (consumed product and donation) is connected with several advantages for the demander. On the one hand, the purchaser's social commitment is expressed by the buying process itself as well as the following consumption. On the other hand, this kind of donation does not cause additional transaction costs for the consumer (like expenses for information or transfer), as would have to be taken in case of a separate purchase of a donation to a relief organisation. The willingness to buy such a commodity bundle can be put down to the fact that the amount of donation linked with the purchase is relatively small. This is a so-called low-cost situation in which ethic behaviour becomes more likely. According to that, KIRCHGÄSSNER (1992, 1996) points out that the lower the costs for moral and ethic behaviour, the higher is the willingness to act in such a way.[16] In opposition to a general donation which is supposed to help many different people, the consumption of fairly-traded products is linked to a more concrete group of recipients. The willingness to help is raised by the knowledge about which group of people can be helped through the buying decision.

14 This linking effect has been used in the campaigns of the Krombacher Brewery and of the football club First F.C. St. Pauli. The concept of the Krombacher initiative is to save one square meter of the Rain Forest from being destroyed by buying one crate of beer. In 2003, landlords in the St Pauli district of the city of Hamburg raised the beer price by 50 cents in order to save the First FC St Pauli Football Club from going bankrupt.

15 For the economic theory of product bundling cf. i.a. Pyndick and Rubinfeld (2001), Varian (2003).

16 About moral and ethic behaviour in low-cost situations cf. (i.a.) Kirchgässner (1992, 1996), Kirchgässner and Pommerehne (1993).

3.2. Features of Goods and Sources of Information

For some consumers, knowing about the conditions of production of food is essential for their buying decision. But often consumers can find out about this property of production only by spending a considerable amount of money even after the consumption of the respective product. From an economic point of view, this can be called credence goods properties.[17] Unlike search properties (e.g. price) or experience properties (e.g. taste), credence goods properties do not even become obvious after the acquisition of the respective product, unless the consumer takes an additional expense in order to determine the credence goods properties or to have them determined.[18] Partly, consumers show an additional willingness to pay for various food products if it is clear to them that those are produced under socially acceptable conditions in the producing countries. But consumers who would like to take into account the production conditions in the producing countries when they decide for a product suffer from an information deficit because the production conditions are not apparent in the product. However, suppliers usually have information about the production conditions. Between both sides of the market there is an information deficit, which can be used by the suppliers. Since demanders cannot make out the production conditions or can do so only by taking an additional expense, suppliers who do not observe social standards in production are encouraged to sell their goods at the price of a high-quality product. If demanders realise the suppliers' opportunity to use this information deficit in their favour, they may possibly abstain from buying food produced in high quality in order not to pay a high price for a product which after all does not contain the promised features, e.g. social production conditions. If the supplier does not succeed in convincing the demanders of the supplied food products' high production quality, the market segment for food produced under observation of certain social production conditions can collapse. Suppliers of goods produced in high quality do not succeed on the market although on the demanders' side there is a willingness to pay for food products produced according to certain social standards. The result could be a process of negative selection, as described by AKERLOF (1970) who uses the example of the second-hand car market[19].

Suppliers have different opportunities to make clear the products' divergence with respect to their production conditions. As described above, distribution via particular chain stores ('third world shops') as well as social trademarks ('social labels') can be considered as signalling instruments. The choice of a particular distribution channel and a social label provides consumers with additional information, which can be taken into account by the consumers when they decide for a product.[20] In case of distribution via third world shops the sales organisation's reputation serves as a guarantee for the correctness of the production conditions which have been promised. Such a mechanism of reputation can work if the supplier has an interest in the demanders repeating their purchase and if establishing a long-term business relation promises higher profits than the short-term benefit from an asymmetry

17 Cf. Darby and Karny (1973).

18 Cf. Haucap and Schmidt (2002), p. 289.

19 Although Akerlof illustrates the process of negative selection with experience goods instead of credence goods, the problems of adverse selection are rather to be expected to become more evident in goods with credence goods properties.

20 Cf. Raynolds (2002), p. 415.

of information.[21] If social trademarks are used, the institutions giving out the certificates vouch for the observation of the (social) standards established by them.

By the indication of production conditions and their surveillance through the relevant institutions the accumulation of useless measurements can be avoided. If demanders whose buying decision is greatly influenced by the way of production had to check the production conditions for themselves, it would cause enormous measurement costs. As emphasized by BARZEL (1982), it is not reasonable for demanders to carry out activities of inspection and measurement after a certain diversification of production quality (in this case of production conditions). If a demander takes measurements and decides against buying the respective product, the same product will presumably be checked by a second consumer. If the second consumer decides against the purchase as well, a third demander will possibly take an expense for measurement and inspection, etc. The result is an accumulation of search and measurement expenses, which can be counterbalanced if the producers or the institutions giving out the certificates indicate the production conditions in a way which is evident to the consumer.[22]

3.3. Social Labels and Price Discrimination

By social trademarks or distribution via third world shops the different production conditions are indicated. This provides the consumer with additional information about the production of the respective item. This indication causes the demanders' different preferences and different willingness to pay to become obvious. Social labels thus allow a self-selection of demanders. Furthermore, it permits the demanders to be divided into different groups of consumers. Segmenting demanders in different groups permits a third-degree price discrimination, i.e. consumers showing different preferences and a willingness to pay different prices for the product which is relatively homogeneous with respect to functional utility.[23] If consumers can be divided in groups with different demand patterns, suppliers will set up their prices in such a way that the group of demanders with the more elastic demand curve pays a lower price than the group of demanders featuring a less elastic demand with regard to price. Suppliers who have a certain amount of market power can demand different prices for an almost identical product from these consumer groups. Transferability of the products between the two markets, i.e. distribution of traditionally produced goods as Fair Trade products, is prevented by the application of different social labels. From the economic point of view, the group of consumers of fairly-traded products forms a club, which can be joined voluntarily by demanders who are willing to pay for the conversion to more social

---

21 For construction of reputation cp. (i.a.) Klein (1985), Telser (1980), Williamson (1975).

22 See Erlei, Esche and Sauerland (1999), p 103.

23 The price of coffee with a Fair Trade label is about 0.5 to 1 Euro (per 100 gram) above the price of conventionally traded coffee. That means that the price of one packet of coffee (500 g) with a Fair Trade label is about 7.50 Euros whereas the price of a packet of traditionally traded coffee is about between 2.50 and 3.50 Euros.

methods of production. The income generated by such a group (of consumers) can be used for achieving the aims envisaged[24].

3.4. Costs of Fundraising and Redistribution

As long as the willingness to pay of some consumers (those who are sensitized for social production) is high enough to cover the transaction costs for supervision and realisation of such quality labels, they can exist. Naturally, the transfer sum available for redistribution is diminished by the Fair Trade organisations' expenses for administration, inspection and marketing. If one wants to assess the ability of the Fair Trade organisations to achieve the envisaged social aims, it is interesting to ask to which extent the expenses for administration, inspection and marketing contribute to the transfer volume of the individual relief organisation. Unfortunately, there are only very few figures available for this assessment. Hence, in the following there will be only a selective comparison (with respect to this criterion) between two relief organisations.

One of the most successful Fair Trade Labelling initiatives, the Max Havelaar Foundation in Switzerland, declares in its annual report of the year 2002 that the transfer volume generated by Fair Trade amounted to a total of 27.5 millions Swiss francs. According to the foundation's account, the proceeds are about 30 per cent higher than the proceeds which producers would have recived for their products in conventional trade. At the same time the figures of the Max Havelaar Foundation's profit and loss account show that in order to acquire these additional receipts of 8.25 million Swiss francs (compared to conventional distribution) expenses amounting to 2.8 million francs were spent on staff, administration, marketing and inspection.[25] Therefore, the expenses for organisation and administration of the foundation amount to about 34 per cent of the generated transfer volume. Compared to that, the annual report of the year 2008 of the popular German relief organisation *Brot für die Welt* documents that the organisation has spent 9,6 million Euros of the total subsidies amounting to 67,1 million Euros on project support, public relations work, advertising and administration. This means that resources to the amount of 57,5 million Euros were available for the concrete relief actions. Organisation and administration expenses therefore have a share of 14,3 per cent in the available redistribution volume of the *Brot für die Welt* Foundation.[26] A comparison of the results shows that, in relation to other relief activities, the contribution of staff and inspection costs to the total redistribution volume of Fair Trade initiatives is relatively high. Yet the share of administration expenses in the gross transfer

---

24 Literature provides various definitions of the term "club", depending on the respective characteristics of the club on which the analysis focuses. „In an attempt to provide a unifying definition, we define a club as a voluntary group deriving mutual benefit from sharing one or more of the following: production costs, the members' characteristics, or a good characterized by excludable benefits" Sandler/Tschirhart (1980), p. 1482. Ng (1973:297) had already pointed out that consumption of club goods is always voluntary whereas pure public goods can be 'non-rejectable' concerning their consumption. For the definition of economic clubs cf. Berglas (1976) and Buchanan (1965).

25 Cf. Max Havelaar Stiftung (2002).

26 Cf. Brot für die Welt (2002). These figures should be interpreted carefully because the annual reports of both foundations only give a rough sketch of the expenditure of the money. A classification into the distinct cost categories is only done roughly.

volume is to a large extent determined by the type of aid.[27] The more specific supporting measures for particular recipient groups are, the more activities of preparation and inspection are required. A lower transfer sum can possibly bring about greater achievements than higher subsidies if specific activities of preparation and inspection are taken to ensure that the subsidies actually reach the groups of people which are in need of help. Therefore the results which have been determined cannot entirely answer the question whether the separate acquisition of a consumer item and a donation can be superior to the product bundle consisting of both components (Fair Trade concept) with regard to the best possible support of the living conditions in developing countries. The relief organisations' statements and the figures available are not sufficient for such an assessment.[28]

4. Effects of Fair Trade on the Producers' Side

4.1. The Individual Rational Choice of a Producer

Let us now turn to the individual rational choice of a producer in a developing country and investigate under which circumstances the participation in Fair Trade is sensible for a producer. Producers have to decide if they produce their goods in a traditional way and sell them at world market price or if they try to observe certain social standards of production and distribute a part of their overall produce at the fair price which is promised by Fair Trade organisations.[29]

4.2. Traditional Production and Distribution on World Market Terms

In order to depict the rational choice of a producer as simply as possible, let us assume that the producer's natural conditions enable him to produce a certain fixed quantity of goods x (with $x > 0$). The costs per unit of quantity produced in a conventional way shall be termed c_w. The world market price per unit of quantity shall be called p_w. Under these assumptions, the profit function of traditional production and distribution at world market conditions turns out to be:

27 In interpreting these results, one should consider the fact that the share of administration costs in the funds available for humanitarian activities will presumably sink if the transfer volume increases because in all probability effects of scale can be realised.

28 The varying efficiency of the different redistribution mechanisms only becomes a problem if non-efficient transfer systems supersede efficient ones while the same groups of recipients are supported. If, on the other hand, new groups of donators are attracted by innovative donation concepts, the redistribution proves to be a Pareto-improvement both for the donator and for the recipient of the donation.

29 A farmer cannot only decide for or against observing social standards in cultivation but also about cultivation itself. For instance, if the price for coffee at the world market decreases, the cultivation of other plants (e.g. of coca) becomes more attractive. Therefore, low coffee prices can cause an increased supply of coca. Cf. Gerster (2001).

$$\Pi_w = (p_w - c_w)x \tag{1}$$

The world market price is determined by global demand and supply. The production costs which have to be taken by the individual producer depend very much on climatic conditions of the respective region, which is particularly true for food production. As a result, the economic bonds[30] of the individual producers partly vary to a large extent. Above, all, harvest-destroying natural disasters can cause situations which threaten the existence of producers since the world market price does not respond to local weather conditions in small regions. Let us now turn to a producer's profit situation if he participates in Fair Trade.

4.3. Social Production and Distribution under Fair Trade Conditions

Let us again suppose that a producer can produce the quantity of goods x. As explained above, producers are guaranteed a fixed sales price (p_f) above the world market price (p_w) in case of participation in Fair Trade. Yet the observation of the social and partly ecological standards demanded by Fair Trade organisations causes some expense. In order to meet the criteria for participation in Fair Trade, the producers have to abstain from forced labour and child labour, keep to internationally obtained employment protection guidelines, and organise cultivation and methods of processing in a non-polluting way.[31] Following these demands of the Fair Trade organisations, let us assume that a producer who decides for the participation in Fair Trade has to take expenses to the amount of c_f for the production of one unit of quantity. These costs for production per unit of quantity are higher than with a traditional way of production because the social standards mentioned above have to be met. It is to be set that $c_f > c_w > 0$.

In reality one can observe that (peasant) producers in developing countries can distribute not the entire quantity of goods, but only a part of it, at the fair price p_f. This is due to the fact that the available transfer sum is determined by the demanders' behaviour and does not suffice to buy the entire quantity of goods produced under Fair Trade conditions at present. The different Fair Trade organisations try to buy from the peasants as much of their produce as possible but the quantity of purchased items is in the end determined by the demand for fairly-traded goods in developed countries. At the moment, producers can only manage to distribute 25 to 50 per cent of the produced quantity of most of their goods at Fair Trade conditions. The rest of the produced quantity they have to sell via conventional distribution channels at world market prices.[32] To put these facts into a formula, let us assume that the part of the overall quantity of production which can be distributed at p_f amounts to π, with $\pi \in [0,1]$. The producers can dispose of the remaining part of the production ($1-\pi$) only at the lower world market price p_w. The revenue (R) from the distribution of the total amount therefore is a function of π:

30 For the definition of the term economic bond cf. Pindyck/ Rubinfeld (2001).

31 For an illustration of Fair Trade criteria see Kunz (1999) and www.transfair.org.

32 Cf. Liebrich (2002), p. 35

$$R_f = (\pi p_f + (1-\pi) p_w) x \quad (2)$$

When social production conditions are observed, the production costs amount to c_f per unit because the Fair Trade organisations demand the entire quantity of production to be produced according to fair production conditions even if only a part of the produced goods can be distributed at Fair Trade conditions.[33] Taking into account these production costs, the profit function of a producer participating in Fair Trade turns out to be:

$$\Pi_f = (\pi p_f + (1-\pi) p_w - c_f) x \quad (3)$$

As can be seen in equations (2) and (3), in case of participation in Fair Trade both the revenue and the profit of a producer depend on which part (π) of the overall produced quantity he can distribute at the fair price p_f.[34] From the assumption $p_f > p_w$ can be deduced that the higher π is, the higher is the producer's profit.

4.4. A Comparison of Profit Situations: Traditional Trade and Fair Trade

Agricultural producers in developing countries have to weigh up whether they profit from conversion of their method of production and participation in Fair Trade. Due to the increased production costs participation in Fair Trade is not always rational for a producer. As mentioned above, it is necessary for the participation in the Fair Trade concept to produce the entire quantity of goods according to the required social standards, which necessarily causes the production costs to rise. But in which case does the conversion of the method of production repay? A producer is undecided whether to join the fair or traditional trade concept if the profit from the distribution at world market price (Π_w) is equal to the profits from distribution at Fair Trade conditions (Π_f). With the help of the profit functions from (1) and (3), the case of profits in fair and in conventional trade being equal can be indicated:

$$(p_w - c_w) x = (\pi p_f + (1-\pi) p_w - c_f) x \quad (4)$$

$$\pi^* = \frac{c_f - c_w}{p_f - p_w} \quad (5)$$

33 The TransFair social label is only given to those producers who produce their goods in peasant cooperatives and are not involved in extensive plantation economy.

34 In market economies (enterprise systems) the market price has an informational, a coordinating and a sanctioning function. Prices which result from exchange processes are storage media to which the individual agents can orientate their actions. Because the Fair Trade organisations provide the producers with a long-term guarantee for the sales price, the price loses in part its function as an informational device. Yet the world market price does not entirely lose its function as a source of information. It is still available as guideline for demanders as well as producers, and producers distribute only a part of the goods produced at Fair Trade conditions.

If a producer can sell the share π^* of his overall production at the price p_f, the profits per unit of quantity in traditional trade correspond exactly to the profits per unit of quantity in Fair Trade. The following graphic illustrates the considerations which have been proposed:

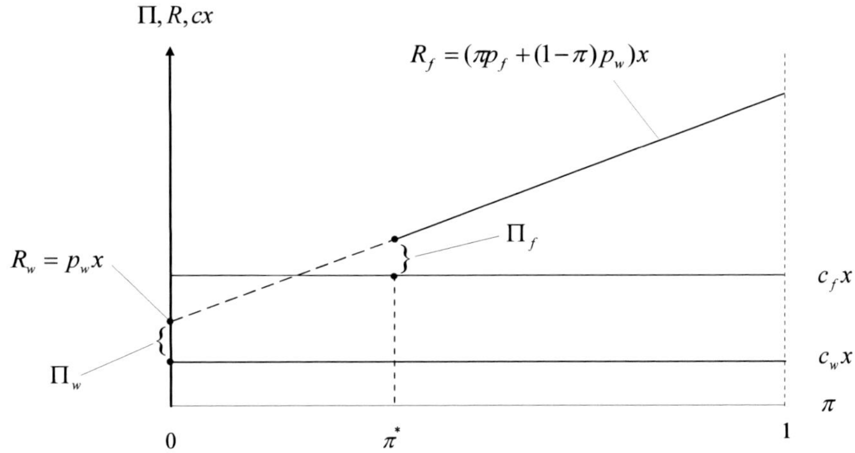

Figure 1. Profits in fair trade and conventional trade

If the profits of participants of Fair Trade are to exceed the profits from conventional trade, the part of the overall production which can be distributed at Fair Trade conditions must be above π^*. In such a situation it is reasonable for a producer to change from conventional methods of production to a more social method of production[35]. If $\pi < \pi^*$, traditional production and sale at world market condition promises higher profits. Unless producers sell at least a share of π^* of their production at the higher fair trade price p_f, there is no (economic) incentive for changing their production methods. If they can sell less than π^* of their production at fair trade conditions, the change to more expensive (social) production methods is not worth while.

As equation (5) illustrates, the part of the overall production which makes the production at Fair Trade conditions appear lucrative (π^*) has to be the higher, the higher are the costs for the social standards which are required (c_f). On the other hand, the sales quota necessary for the conversion to more social production method is the lower, the higher above the world market price is the guaranteed fair price (p_f). If the world market prices for agricultural products fall, as has often been the case in the last decade, profits of the distribution at world market conditions fall *ceteris paribus.* This brings about a stronger incentive for producers to join the Fair Trade register.

35 We assume that producers behave risk neutral. If they behave risk-averse, producers will remain in traditional (unsocial) production structures for a longer time because the high-risk change of production methods is only realised if the risk is compensated appropriately by higher proceeds. For reasons of simplicity we will continue to assume that there are risk neutral economic agents

4.5. Membership and Sales Quotas in the Fair Trade Club

The sales quota which can be distributed by an individual producer at the fair price p_f is determined on the one hand by the overall-demanded quantity of fairly-traded goods. On the other hand, it is determined by the total number of suppliers in the Fair Trade register and their quantity of production. Provided that the demanded quantity does not increase, the transfer sum set for distribution is fixed and an influx of new suppliers must lead to a decrease of the sales quota (π). The following quote shows that such a development can actually be observed in reality, especially with respect to coffee, the first Fair Trade product:

> "The strong growth of Fair Trade coffee sales in Europe over the last decade is tempered by the fact that demand is still insufficient to absorb the total supply from the 550,000 small farmers on the Fair Trade Register. In fact, only about half of the total production of these groups is presently sold at Fair Trade terms, due to the limited market. Nevertheless, the producers themselves have rejected the option of closing the Register to newcomers or imposing quotas." (RICE und MCLEAN 1999, p. 79)

One reason for not closing the club could be found in the Fair Trade organisations' aim to bring about a change for social production methods instead of providing additional profits for those who were the first to realise the advantages of fair trade. As will be shown in the following, the aim of fair trade organisations can also be achieved by non-regulated club access since a number of members will be reached which provides all producers in the Fair Trade register with proceeds sufficient to pay for the additional costs of social production.

But which consequences are to be expected from the admission of new producers in the Fair Trade club and how long is there a stimulus to join this club? As shown by the model considerations above, the profits from the participation in Fair Trade are only higher than the profits which can be gained in conventional trade if $\pi > \pi^*$. Consequently, new producers will only join the Fair Trade club as long as $\pi > \pi^*$. If there are already so many producers participating in the Fair Trade concept that all suppliers can only distribute the part π^* of their overall production at the high sales price, the influx of new producers will recede. Provided that $\pi < \pi^*$, producers will change from Fair Trade to conventional trade because in a situation like that the production at costs of c_w and the revenue gained by distribution at world market price promises a higher profit than participation in the Fair Trade concept. These reflections show that distribution at Fair Trade conditions is only reasonable for a producer if $\pi \geq \pi^*$. Otherwise, the expenditure necessary for the conversion of production by observing the social production standards causes a considerable loss of profits. A rational producer will refrain from doing so. This analysis shows that sharing the risk between producers and the Fair Trade initiatives may be an important issue. Since the organisations are the better-informed agents, it seems reasonable that they guarantee a certain level of Fair Trade consumption in the donator countries. The low redistribution efficiency (chapter 3.4.) suggests that parts of the donation revenue are used to assure Fair Trade consumption at the current level in the future.

Especially producers who have a small amount of c_f - c_w, i.e. who have no great additional costs caused by the change in production, will adjust production to more social methods. Because c_f - c_w will regionally differ, Fair Trade activities are likely to favour producers which are closer to fulfilling the social standards required by the initiatives. Thus, the spatial distribution of the gains of Fair Trade is determined.

The graph shows that π^* is endogenously determined by the profit expectations in conventional trade. Therefore, a fall in world prices of commonly produced agricultural products will lower the decisive share of output which is sold at the higher price of labelled products. This implies that the establishment of Fair Trade institutions causes a leverage effect on the overall share of social production standards with respect to all production activities. Not only the fairly-traded amount of production, but the rest of a peasant producer's cultivated land as well becomes a place of socially acceptable production. This is a possible source of excess utility for the (well-informed) consumer of Fair Trade products. The introduction of social standards for production activities which is not awarded but claimed by the Fair Trade organisations is not reflected in the redistribution efficiency statistics quoted in chapter 3.4. However, this extension of social production through the described leverage effect is probably to be considered as beneficial and desirable by the consumers of Fair Trade products.

According to the considerations which were just presented, it is only lucrative for a part of all producers to participate in the Fair Trade concept whereas for other producers joining the club does not repay after a certain number of participants has been attained. A balance will set in which is marked by

a) a certain number of producers fulfilling the social standards envisaged by the Fair Trade organisations, and
b) these producers distributing a part of their overall production to the amount of π^* at Fair Trade conditions.

Provided that such a balanced sales quota of π^* can develop, the profits from Fair Trade are just sufficient to cover the additional costs which are required by the observance of the envisaged social aims.[36] Due to the restricted transfer sum the Fair Trade organisations have to ask themselves which social criteria shall be required for the participation in Fair Trade: The lower are the required social standards, and therefore the costs for meeting them, the more producers can be given the opportunity to keep to these social standards in production. The more extensive and cost-intensive the required social criteria are, the fewer producers will profit from these social standards and be able to produce their goods under improved working conditions.

36 Especially producers from Central and South America have already realised the advantages which are offered to them by the fair distribution of their products at an early time. At present, the majority of Fair Trade producers are situated in this continent whereas only a few Fair Trade producers are registered in Africa and Asia (Rice and McLean 1999, p. 58). Provided that the demanded quantity of fairly-traded products does not increase, it is to be supposed that Asian and African producers will not join the Fair Trade concept because of the number of participants being already too high and the quotas (π) being too low to become a member.

4.6. Voluntary Social Standards and International Trade

In order to guarantee social standards in as many countries as possible, some developed countries demand the incorporation of socio-political regulations in the WTO. Yet these endeavours are rejected by numerous developing countries.[37] The Fair Trade initiatives based on voluntariness are interesting concepts for the accomplishment of more humane methods of production. In opposition to social minimum standards they try to provide a positive incentive. In comparison to internationally obligatory social standards these initiatives - unlike legally determined minimum social standards for the production - have the advantage to represent the demanders' actual willingness to pay for more humane methods of production. Thus the Fair Trade initiatives draw some attention on the consumers' interest in more humane or social working and living conditions of producers in developing countries. This might provoke interesting but probably misleading questions about the moral character of trade between developing and industrial countries. The bare existence of an alternative to traditionally produced export goods of third world origin is capable to make consumers aware of their own willingness to act charitably and might be a source of compunction.

Legal minimum standards require all demanders to take higher expenses for production and inspection. With the Fair Trade concept, which is based on voluntariness, only those demanders who actually have the willingness to pay for social methods of production and express this willingness in their purchase pattern are charged to take additional expenses. Furthermore, the concept of Fair Trade does not trigger protectionist effects which can occur with the regulation of compulsory social standards.[38] Whereas establishing social minimum standards and introducing punishment for not obeying to them can cause a protection effect,[39] the Fair Trade principle is based on voluntariness and is not of an obliging nature. This is the reason why the WTO does not attack it as a protectionist trade barrier.[40] Producers still have the opportunity to distribute their products in a traditional way. They are not excluded from participation in international trade.

The incentives for reorganizing the production process which are brought about by Fair Trade initiatives have two effects which will not be analysed further in this paper. On the one hand, they are to be expected to have learning effects on producers. As a consequence, the difference between the costs of the two production methods will be diminished in the course of time. On the other hand, the factor input relation will change towards a more capital-intensive production method. Since Fair Trade initiatives reward producers who take higher expenses for the labour utilized in the production, a higher labour productivity is subsidized. This results in an increased capital equipment of jobs.

---

37 Cf. Schneuwly (2003) as well as Busse and Grossmann (2003). For an overview of the arguments of the developing countries against the regulation of compulsary social standards see Slazar-Xirinachs (1999).

38 For a more extensive discussion of the effects of social labels on trade see Zadek, Simon, Lingayah, Sanjiv, Forstater, Maya (1998),p. 69ff..

39 The introduction of internationally binding social standards holds the danger that such minimum standards develop protectionist effects and are exploited by lobbyists. Cf. Freeman (1994).

40 Cf. Zadek, Simon, Lingayah, Sanjiv, Forstater, Maya (1998), p. 71.

5. Summary

By Fair Trade initiatives consumers are given the possibility to signal their willingness to pay for a way of production of various goods which is characterized by sustainability as well as social justice. As a result, Fair Trade participants experience Pareto-improvement because both demanders and producers profit by the voluntary exchanges. Consumers of Fair Trade products are willing to pay more than the current market price in order to achieve an additional benefit. On the other hand, producers can use additional receipts for the conversion to more humane methods of production, which is in the producers' interest as well. The construction of Fair Trade itself unites two processes: fundraising and redistribution. Producers become personified, which establishes a bond between consumers and producers. This connection stimulates the tendency to help, but at the same time offers an opportunity for active support. Probably there are more efficient mechanisms of redistribution than the rather expensive Fair Trade concept. Nevertheless, the combination of fundraising and direct redistribution seems to work. The aims envisaged by the Fair Trade organisations can be achieved by the mechanism which is used. At the moment, the demand for Fair Traded products does only form a restricted part of the overall demand for products so that the transfer sum available for distribution is relatively small. This means that only a few producers can be given the opportunity to accomplish more social methods of production and a more social way of life. From the economic point of view, an extension of such Fair Trade initiatives is to be appreciated because it provides a device based on voluntariness which helps to counteract the social problems caused or revealed by international trade.

References

Akerlof, George A. (1970), The Market for „Lemons“: Quality Uncertainty and the Market Mechanism, *Quarterly Journal of Economics* 84 (3), pp. 488-500.

Antle, J.M. (1999), The new economics for agriculture, *American Journal of Agricultural Economics* 81, No. 5, pp. 993-1010.

Barzel, Y. (1982), Measurement cost and the Organization of Markets, *Journal of Law and Economics* 25 (1), pp. 27-48.

Becker, Ursula (1995), *Entwicklung und Organisation des Hanseatischen Kaffeehandels im 19.* und 20. Jahrhundert, Köln.

Berglas, E. (1976), "On the Theory of Clubs", *American Economic Review,* Vol. 66, pp. 116–121.

Buchanan, J. M. (1965), "*An Economic Theory of Clubs", Economica,* Vol. 32, pp. 1–14.

Brot für die Welt (2002), Jahresbericht 2008, Internet: *http://www.brot-fuer-die-welt.de/downloads/ueber-uns/Jahresbericht_2008.pdf,* (December, 09, 2009).

Busse, Mathias und Harald Grossmann (2003), Handelsbezogene Aspekte sozialer Mindeststandards, *Wirtschaftsdienst* 83 (2), pp. 125-129.

Darby, michael und Edi Karni (1973), Free Competition and the Optimal Amount of Fraud, *Journal of Law and Economics* 16 (1), pp. 67-88.

Erlei, Mathias, Martin Leschke und Dirk Sauerland (1999), *Neue Institutionenökonomik*, Stuttgart, Schäffer-Poeschel.

Freeman, R. (1994), A Hard-headed Look at Labour Standards, in: W. Sengenberger und D. Cambell (eds.), *International Labour Standards and Economic Interdependence, Genf*, pp. 117-157.

Gerster, Richard (2001), Globalisierung und Gerechtigkeit, Bern.

Haucap, Justus and Helmmar Schmidt (2002), *Kennzeichnungspflicht für genetisch veränderte Lebensmittel: Eine ökonomische Analyse, Zeitschrift für Wirtschaftspolitik* 51 (3), pp. 287-316.

International Coffee Organization (2004), Internet: *http://www.ico.org* (April, 17, 2006).

Kirchgässner, Gebhard (1996), *Bemerkungen zur Minimalmoral, Zeitschrift für Wirtschafts- und Sozialwissenschaften* 116 (2), pp. 223-251.

Kirchgässner, Gebhard and Werner W. Pommerehne (1993), *Low-Cost Decisions as a Challenge to Public Choice, Public Choice* 77 (1), pp. 107-115.

Kirchgässner, Gebhard (1992), Towards a Theory of Low-Cost Decisions, *European Journal of Political Economy* 8 (2), pp. 305-320.

Klein, Benjamin (1985), Self enforcing contracts, *Journal of Institutional and Theoretical Economics* (JITE) 141, pp. 594-600.

Kommission der Europäischen Gemeinschaften (1999), *Mitteilung der Kommission an den Rat über "fairen Handel"*, KOM(1999) 619, Brüssel.

Kunz, Martin (1999), Fair Trade im Vergleich mit anderen Bemühungen, Arbeitsbedingungen in der globalen Wirtschaft zu verbessern, Internet: *http://www.fairtrade.de* (April, 17, 2006).

Liebrich, Andreas (2002), *Standards und Labels - Das Beispiel des fairen Handels,* discussion paper No. 95, Institut für Wirtschaftsethik der Universität St. Gallen.

Max Havelaar Stiftung (2002), Fair Trade Jahresbericht 2002, Internet: *www.maxhavelaar.ch/web/havelaar/mainR2.nsf/MHOrderList*?OpenForm&03&MHOrderList (April, 17, 2006).

Morgenbesser, Herbert (1959), *Der internationale Kaffeemarkt, Wirtschaftsdienst* 39, pp. 101-104.

Mummert, Uwe (1995), *Informelle Institutionen in ökonomischen Transformationsprozessen*, Baden-Baden.

Ng, Y.-K. (1973), "The Economic Theory of Clubs: Pareto-Optimality Conditions", *Economica*, Vol. 40, pp. 291–298.

Ottowitz, Taciana Valio (1997), *Verbraucherverhalten bei fair gehandeltem Kaffee:* Ergebnisse theoretischer Ableitungen und empirischer Untersuchungen, Frankfurt am Main.

Pindyck, Robert S. and Daniel L. Rubinfeld (2001), *Microeconomics*, 5th Ed., New Jersey.

Piepel, Klaus (2000), *Sozialsiegel und Verhaltenskodizes - Eine Standortbestimmung, Aachen: Misereor Medienproduktion und Verlagsgesellschaft.*

Raynolds, Laura T. (2002), Consumer/Producer Links in Fair Trade Coffee Networks, *Sociologia Ruralis* 42 (4), pp. 404-424.

Rice, Paul D. and Jennifer McLean (1999), Sustainable Coffee at the Crossroads, A White Paper Prepared For: The Consumer's Choice Council, Internet: *http://www.consumerscouncil.org/coffee/coffeebook/coffeebook.htm* (April, 17, 2006).

Salazar-Xirinachs, Josè M. (1999), The Trade-Labor Nexus: Developing Countries Perspective, Internet: *http:://www.sice.oas.org/tunit/STAFF_ARTICLE/ jmsx_labor1.asp* (April, 17, 2006).

Sandler, T., Tschirhart, J. (1980), "The Economic Theory of Clubs: An Evaluative Survey", *Journal of Economic Literature*, Vol. 18, pp. 1481–1521.

Schneuwly, Phillippe (2003), Sind Handelssanktionen ein geeignetes Mittel zur Durchsetzung von Arbeitsnormen? Eine Untersuchung der Wirksamkeit der Sozialklausel im US GSP, *Aussenwirtschaft* 58 (1) , pp. 121-144.

Telser, Lester G. (1980), A Theory of Self-enforcing Agreements, *Journal of Business* 53 (1), pp. 27-44.

Varian, Hal R. (2003), *Intermediate microeconomics : a modern approach,* 6. ed., internat. student ed. - New York.

Weißmann, Norbert and Hermann-Josef Wirtz (2001), *Im Zeichen der Nachhaltigkeit - Verknüpfung von Öko- & Fair-Trade-Initiativen, Verein zur Förderung von Gerechtigkeit im Welthandel* (eds.), Wuppertal.

Welford, Richard, Julia Meaton and William Young (2003), Fair trade as a strategy for international competitiveness, *International Journal of Sustainable Development World Ecology* (10), pp. 1-13.

Williamson, Oliver (1975), *Markets and Hierarchies: Analysis and Antitrust Implications; a study in the economics of internal organization,* New York: Free Press.

Zadek, Simon, Lingayah, Sanjiv and Forstater, Maya (1998), *Social Labels: Tools for Ethical Trade, Final report of the New Economics Foundation,* prepaired for: Directorate-General for Employment, Industrial Relations and Social Affairs Directorate V/D.1 of the European Commission.

In: Agricultural Production
Editors: Felix C. Wager, pp. 201-216
ISBN: 978-1-61668-695-6

Chapter 10

IMPROVEMENT OF PLANT RESISTANCE TO PATHOGENS AND INSECTS BY THE USE OF GENES OF PROTEASE INHIBITORS

Y.E. Dunaevsky*, E.N. Elpidina and M.A. Belozersky
A.N. Belozersky Institute of Physicochemical Biology, Moscow State University, Moscow, Russia

ABSTRACT

This review concerns the possibility of using plant inhibitors of proteolytic enzymes to improve plant resistance to insects and phytopathogens. The main argument in favor of this approach is that protease inhibitors are widespread in plant tissues and highly active with respect to various proteases of insects, bacteria, and fungi. Results obtained for the last years point to perspective of genetic engineering approach. Main deficiencies and advantages of this approach and ways to improve its efficiency are discussed. Possibility of obtaining tobacco and potato plants resistant to bacterial infections by introduction of only one gene of serine proteinase inhibitor BWI-1a from buckwheat seeds through agrobacterial transformation of leaf disks has been demonstrated. It is noted that the discussed approach to increase plant tolerance to insects and pathogens has a number of advantages in comparison with traditional techniques and is ecologically pure.

Keywords: agrobacterial transformation, filamentous fungi, insects, phytopathogens, proteinase inhibitor, proteinase inhibitor gene, transgenic plants

* Corresponding author. E-mail: dun@belozersky.msu.ru, A.N. Belozersky Institute of Physicochemical Biology, Moscow State University, Moscow, 119992 Russia

INTRODUCTION

An increase in agricultural production and the improvement of crop preservation are among the main challenges standing before the continuously growing human population, which is expected to reach or even exceed ten billion by the mid-21st century. Current approaches to these problems rely mostly on chemicals: fertilizers and protectants (insecticides, fungicides, herbicides, etc.). The wide use of synthetic compounds creates an excessively high pressure on the environment and destabilizes the existing ecological balance. Moreover, standing the huge expenditures (more than ten billion dollars annually) on production of synthetic chemicals, the annual loss reaches 37% of the total crop and its cost is estimated at 300 billion dollars; insect pests are responsible for 13% of the loss [Thomas 1999]. Plant bacterial diseases provoke a great economical loss, causing damage of valuable fruit, essential oil, technical and vegetable plants. Some bacteriosis affect plants on vast territories – for example, "burn" of seminal and fruit cultures, bacterial senescence of maize. Black bacteriosis of wheat results in loss of 15-60% of crop, and cotton hommosis – of 60 %. Strong early affection of tomato plants with tomato cancer may lead to loss of 70-80% of yield. Although more specific and less stable protectants are continuously developed to reduce the ecological hazard created by using them, the consumption of synthetic chemicals has started to decrease (mostly in Europe). Hence, it is necessary to design new technologies in order to reduce the use of synthetic compounds without increasing the loss of crops or to even decrease the loss.

In addition to special agrotechnical measures and breeding with higher resistance to various pests, genetic engineering can be used to solve this problem and to obtain transgenic plants resistant to diseases and insects. Apart from some minor, poorly studied possibilities, there are two main approaches to plant transformation with genes capable of controlling the development and abundance of pests. One involves genes that are alien to plants: the *Bacillus thuringiensis* endotoxin genes and, to a lesser extent, the genes for some microbial extracellular proteins, scorpion and spider neurotoxins, and bacterial enzymes. The other approach takes advantage of the plant genes that code for defense proteins, resulting from prolonged coevolution of plants and pests. These proteins include various hydrolase inhibitors, lectins, and some plant enzymes. Each of the two approaches has its advantages and drawbacks. Yet gene transfer from one to another plant is more appealing, as it is simpler and safer than transfer of foreign genes.

More than one hundred million years of coevolution have allowed plants to acquire several effective mechanisms protecting them from attack by phytopathogens and pests. The most interesting are mechanisms that depend on plant proteins exerting a toxic or antimetabolic effect on pests. These proteins include hydrolase inhibitors, lectins, ribosome-inactivating proteins, enzymes (chitinase, lipoxygenase, acylhydrolase, etc.), and many others, and each contributes to the generation of a protective barrier in early infection. It is quite natural that protease inhibitors play a protective role, because both insects and pathogenic microorganisms utilize proteolytic enzymes to penetrate host tissues or to feed on them. Plants utilize constitutive and inducible protease inhibitors to prevent pathogen invasion.

In the present review we consider the data suggesting a role of plant protease inhibitors in plant protection from insect pests and the our results of their employment in genetic engineering works aimed at constructing cultures with a higher resistance to pests and phytopathogens.

PROTEINASE INHIBITORS IN PLANT DEFENSE AGAINST BIOTIC STRESS

The above inhibitors are proteins that interact with proteases and suppress their proteolytic activity. Such inhibitors are widespread in plant tissues and specific for one of the four types of proteases (serine, cysteine, and aspartyl proteases and metalloproteases). Inhibitors of serine and cysteine proteases are of most interest among them, because these inhibitors prevail in plants and, on the other hand, their target enzymes are most common in insects and phytopathogens.

Serine and cysteine protease inhibitors are mostly small proteins, 6-25 kDa in molecular weight. Their content in seeds reaches 4-10% of the total water-soluble protein. The protective role of such inhibitors is supported by their preferential association with seeds, which are most attractive to various pests and most vulnerable because of their metabolic inactivity and the consequent lack of an active defense system. Moreover, inhibitors are highly stable under unfavorable conditions (extreme pH, higher temperatures, etc.). Another argument in favor of the protective role of protease inhibitors is that inhibitors originating from the same source greatly vary in specificity.

It is noteworthy that most inhibitors exert no effect on cognate seed proteases and highly efficiently inhibit animal, fungal and bacterial proteolytic enzymes as well as proteases of the digestive tract of insect pests.

Not only do the inhibitors affect the above proteases *in vitro,* but they also suppress larval growth and pupation when added to the diet at 2-5%. For instance, the Bowman-Birk and Kunitz soybean trypsin inhibitors, as well as the trypsin inhibitor of the *Vigna* cowpea, considerably reduce the biomass of *Tribolium confusum, T. casteneum,* and *Heliothis virescens* larvae when added to the artificial diet [Ryan 1990; Gatehouse et al. 1993]. Used at a higher (10%) concentration, the Kunitz trypsin inhibitor and potato inhibitor II impair the growth and the physiology of digestion in *H. zea* and *Spodoptera exigua* larvae [Ryan 1990]. The cysteine protease inhibitor oryzacystatin exerts a stable suppressor effect on the viability and development of *Diabrotica undecimpunctata* larvae [Edmonds et al. 1996], which are tolerant of *B. thuringiensis* endotoxins. As components of the artificial diet, serine protease inhibitors, aminopeptidases, and carboxypeptidases synergistically affect the development of the mite *Tyrophagus putrescentiae* [Ortego et al. 2000].

A positive correlation was observed in some cases between the content of protease inhibitors and pest resistance of plants. For instance, a high level of inhibitors is responsible, at least to a certain extent, for the resistance of a *Vigna* cultivar to cowpea weevil *Callosobruchus maculates,* the main pest of *Vigna* [Gatehouse et al. 1992]. A high content of serine protease inhibitors in seeds correlates well with biscuit weevil resistance in durum and common wheat [Konarev 1987]. Moreover, a considerable decrease in resistance to *Manduca sexta* larvae was observed in transgenic tomatoes expressing the prosystemin gene in the antisense orientation and, consequently, producing protease inhibitors at a lower level [Orozco-Cardenas et al. 1993].

Taken together, the above data justify the use of protease inhibitors genes for constructing transgenic plants with a higher pest resistance. The first success in this direction was achieved in 1987 [Hilder et al. 1987]. Constitutive expression of a *Vigna* protease inhibitor gene in transgenic tobacco plants yielded impressive results. Biological testing with cutworm *H. virescens* showed that transgenic plants producing the alien inhibitor to at least 1% of the total soluble protein reduced the growth and survival of larvae as well as the damage area. Similar results were obtained in field experiments with *Helicoverpa zea* [Lecardonnell et al. 1999]. Transgenic plants expressing genes for exogenous serine protease inhibitors have a higher insect resistance as judged from insect survival, insect biomass, and plant damage. Recent studies considered the transfer of the gene for the cysteine protease inhibitor oryzacystatin and yielded potato plants with a higher resistance to Colorado beetle larvae, the main pest of potatoes [Hilder and Boulter 1999].

Table 1. Biotest of plants transformed with protease inhibitors genes.

Plant source (inhibitor)[a]	Transgenic plant	Insect[b]	Effect[c]	Reference
Cowpea (CpTI)	Pigeonpea	*Helicowerpa armigera* (Lep)	+	Lawrence & Koundal 2002
Barley (TI)	Wheat	*Spodoptera littoralis* (Lep)	+	Altpeter et al. 1999
		Melanoplus sanguinipes (Ort)	_	Altpeter et al. 1999
(CMe)	Rice	*Sitophilus oryzae* (Col)	+	Alfonso-Rubi et al. 2003
Mustard (MTI-2)	Tobacco, *Arabidopsis*	*Spodoptera littoralis* (Lep)	+	De Leo et al. 2002
	Tobacco	*Plutella xylostella* (Lep)	+	De Leo et al. 2001
	Arabidopsis	*Mamestra brassicae* (Lep)	+	De Leo et al. 2001
Nicotiana alata (PI)	Pea	*Plutella xylostella* (Lep)	+	Charity et al. 1999
	Tobacco, pea	*Helicowerpa armigera* (Lep)	+	Charity et al. 1999
Rice (OC-1)	Potato	*Leptinotarsa decemlineata* (Col)	_	Cloutier et al. 2000
(OC-II)	Alfalfa	*Phytodecta fornicata* (Col)	+	Ninković et al. 2007
Soybean (TIK)	Tobacco	*Spodoptera litura* (Lep)	+	McManus & Burgess 1999
	Rice	*Nilaparvata lugens* (Hem)	+	Lee et al. 1999
(C-II , PI-IV)	Tobacco, potato	*Spodoptera littoralis* (Lep)	+	Marchetti et al. 2000
(TIK, TIBB)	Sugarcane	*Diatraea saccharalis* (Lep)	+	Falco & Silva-Filho 2003
Solanum americanum (SaPIN2a)	Lettuce	*Trichoplusia ni* (Lep)	+	Xu et al. 2004
	Tobacco	*Spodoptera litura,* (Lep) *Helicoverpa armigera* (Lep)	+	Luo et al. 2009
Arabidopsis thaliana (CPI)	White poplar	*Chrysomela populi* (Col)	+	Delledonne et al. 2001
Tobacco	Tobacco	*Spodoptera litura,* (Lep) *Helicoverpa armigera* (Lep)	+	Srinivasan et al. 2009
Maize (mpi)	Rice	*Chilo suppressalis* (Lep)	+	Vila et al. 2005
Potato (PINII-2x)	Rice	*Chilo suppressalis* (Lep)	+	Bu et al. 2006

[a] OC – oryzacystatine, PI – proteinase inhibitor, TI – trypsin inhibitor, TIBB – Bowman-Birk trypsin inhibitor, TIK – Kunitz trypsin inhibitor

[b] Col - Coleoptera; Hem – Hemiptera; Lep – Lepidoptera; Ort - Orthoptera

[c] + significant differences between transgenic plants and control; - without significant differences

In total, about 20 works reported the construction of various transgenic plants with protease inhibitor genes from various sources over the past decade (Table 1). A positive effect was achieved in about 90% of cases (including field testing): transgenic plants displayed a higher resistance to various insects mainly of the orders Lepidoptera and Coleoptera.

However, for the last years some works appeared indicating on existence of certain limitations of using proteinase inhibitors in transgenic plants. It was found insects that are reared on a medium enriched in an inhibitor or that feed on transgenic plants can adapt to the new conditions and considerably, if not completely, evade the effect of the inhibitor. Such evasion is due to degradation of the inhibitor by another, non-target protease [Michaud et al. 1996; Girard et al. 1998; Mazumdar-Leighton and Broadway 2001a; Mazumdar-Lcighton and Broadway 2001b; Zhu-Salzman et al. 2003]; overproduction of the target protease (such overproduction was observed for a cysteine protease isoform that allows a highly efficient autocatalytic transformation of the latent proenzyme into the active mature form, which cleaves the alimentary inhibitor) [De Leo et al. 1998; Ahn et al. 2004]; or synthesis of a new protease that is tolerant of the exogenous inhibitor and may belong to another class of the family (e.g., a cysteine in place of a serine protease or chymotrypsin and elastase in place of trypsin) [Jongsma et al. 1995; Cloutier et al 2000; Wu et al. 1997; Bouchard et al. 2003]. It is probable that adaptation to a new inhibitor depends on the combined effect of several factors: proteases either of the same type as the target enzyme or of another type are produced to a considerably higher level and become capable, possibly as a result of their joint action, of cleaving the inhibitor used in transgenic plants [Zhu-Salzman et al. 2003].

One should hope that discovery of these limitations will stimulate a search for new means to overcome such adaptation. In particular, expression of insect genes for inhibitors that regulate cognate insect digestive proteases in transgenic plants may provide a specific efficient means against insect pests [Thomas et al. 1995]. A considerable positive effect may be achieved by simultaneous introduction of both serine and cysteine protease inhibitors in transgenic plants: degradation of the inhibitors by tolerant endogenous proteases is less likely in this case. Multidomain inhibitors of cysteine proteases were constructed from several individual inhibitors by genetic engineering techniques, and their genes expressed in potatoes. As a result, the number of thrips *Frankliniella oeeidentalis* larvae on transgenic plants was reduced by 80% and the damage to leaves was minimal [Outchkourov et al. 2004a; Outchkourov et al. 2004b].

Interesting and promising results were obtained with transgenic plants carrying two foreign genes that code for proteins differing in the mechanism of action. In transgenic tobacco, the *Vigna* trypsin inhibitor and pea lectin genes produced an additive protective effect against *Heliothis vireseens* as judged from the decrease in larval biomass and leaf damage [Boulter et al. 1990]. Low level expression of serine protease inhibitors was reported to increase the insecticide activity of *B. thuringiensis* endotoxins [MacIntosh et al. 1990].

An important field of research is seeking for new inhibitors specific for major digestive proteases of insects. Studies of the Colorado beetle digestive complex showed that aspartyl protease starts hydrolysis of the protein component of the food. Hence, expression of a gene for its inhibitor in potato is potentially more effective than expression of the oryzacystatin gene and may play a crucial role in suppressing insect digestion [Brunelle et al. 1999].

The effect of alien protease inhibitors, synthesized in transgenic plants, on the host's metabolic functions has attracted particular interest in recent years. To prevent possible adverse effects of an alien inhibitor, it was proposed that the inhibitor is synthesized as an inactive precursor with a specific site recognized by an insect digestive protease: hydrolysis at this site yields the active inhibitor. Another possible means to utilize the insecticide potential of protease inhibitors is to put expression of a foreign inhibitor gene under the control of a specific inductor(s) associated with an attack by insects [Koiwa et al. 2000].

It should be noted that the direct effect of plant protease inhibitors on pathogenic microorganisms is understood to a far lesser extent than their effect (direct or in transgenic plants) on insects, although fungal and bacterial diseases are among the major biotic stress factors with a dramatic impact on the crop yield. A few studies considered the effect of inhibitors on phytopathogen proteases *in vitro* along with their effect on microorganisms at various developmental stages. For instance, a fungicide-like effect was demonstrated for a serine protease inhibitor from cabbage leaves [Lorito et al. 1994] and a cysteine protease inhibitor from millet seeds [Joshi et al. 1998]. Bowman-Birk inhibitors from beans slowed down hyphae growth and germination of conidium of fungi *Fusarium culmorum, F. solani* and *Botrytis cinerea* [Benken et al. 1976]. Trypsin inhibitor from buckwheat also slowed down spore germination and mycelium growth of phytopathogenic fungi *A. alternata* and *F. oxysporum* [Dunaevsky et al. 1998], and trypsin inhibitor from maize caused a similar effect towards pathogenic fungi *Aspergillus flavus, Asp. parasiticus* and *F. moniliforme* [Chen et al. 1999]. It was demonstrated that trypsin inhibitors from potato tubers suppressed growth and development of pathogen *Phytophthora infestans* [Valueva et al. 1998; Kladnitskaya et al. 1996]. Cysteine proteinase inhibitors can play an important role in interrelations of plants and viruses [Kondo et al. 1992]. It is due to the fact that cysteine proteinases play an active role in protein processing in many viruses. In this connection cysteine proteinases become an attractive target in course of development of efficient techniques against viral diseases both of plants and animals [Gorbalenya and Snijder 1996]. As in the case with other pathogenesis related (PR) proteins [Van Loon 1985], synthesis of proteinase inhibitors was induced in response to infection with phytopathogenic microorganisms. This effect was observed first with tomato plants infected with oomycete *Phytophthora infestans*. At that correlation between increase in the content of trypsin and chymotrypsin inhibitors and plant resistance to pathogen took place [Peng and Black 1976]. Increase in the activity of serine proteinase inhibitor in potato tubers infected with *P. infestans* was also observed [Kladnitskaya et al. 1996]. Similar effect was took place not only for plants of solanaceous family but for plants of other families too. Thus, infection of melon with fungus *Colletotrichum lagenarium* caused increase in the activity of inhibitor reacting with proteinase of the pathogen [Roby et al. 1987]. Similar effect was observed for representatives of cereals. Infection of maize seedlings with fungus *F. moniliforme* resulted in both local and systemic induction of serine proteinase inhibitor belonging to potato proteinase inhibitor I family [Cordero et al. 1994], and formation of cystatine took place in chestnut leaves in response to infection with the fungus *Botrytis cinerea* [Pernas et al. 2000].

However, in spite of the fact that losses of crop caused by phytopathogens are most significant from the quantitative point of view there are only few works reporting on obtaining of transgenic plants, containing proteinase inhibitor gene, with increased resistance to pathogenic microflora [Charity et al. 2005].

USE OF BWI-1A INHIBITOR FROM BUCKWHEAT SEEDS FOR OBTAINING TRANSGENIC PLANTS WITH INCREASED RESISTANCE TO PATHOGENIC MICROFLORA AND INSECTS

In our work one of trypsin inhibitors from buckwheat seeds (BWI-1a) was used for obtaining transgenic plants with increased resistance to pathogenic microflora. This inhibitor was represented well quantitatively and was able to inhibit growth of hyphae and spore germination of fungi of phytopathogenic species *Alternaria alternata* and *Fusarium oxysporum*, grown in Petri dishes or under "hanging drop" conditions.

The structure of the target gene was deduced from the amino acid sequence of buckwheat seed anionic proteinase inhibitor BWI-1a. The 213 bp inhibitor gene (*ISP*) was directly assembled from a number of synthetic polynucleotides by PCR technique.

Agrobacterium tumefaciens A281 strain with pBI 101 vector, containing *ISP* gene under 35S promoter of cauliflower mosaic virus, was used for transformation of aseptic tobacco cv. Samsun (*Nicotiana tabacum* L.) and potato cv. Reserv (*Solanum tuberosum* L.) plants. Cysteine 400 μg/ml was added to cultivation and regeneration media during plant transformation [Olhoft and Somers 2001]. Primary regenerates were obtained from transformed leaf segments on the following regeneration media: 2 μg/ml of 6-benzylaminopurine (6BAP) and 0.2 μg/ml of α-naphtylacetic acid were added to the medium for tobacco and 3 μg/ml of 6BAP - for potato regenerates. In the course of selection of transformants kanamycin sulfate (100-150 μg/ml) and klaforan (250-500 μg/ml) were added to all media to remove agrobacteria.

Analysis of presence and expression of the functional *ISP* gene in the cells of transformed potato and tobacco lines indicated that all transformed lines contained a common fragment of $0.6.10^3$ bp corresponding to the fragment of gene construction. On the whole, predominance of single insertions was observed what corresponds with literature data on insertion of gene constructions in the course of T-DNA-mediated agrobacterial transformation.

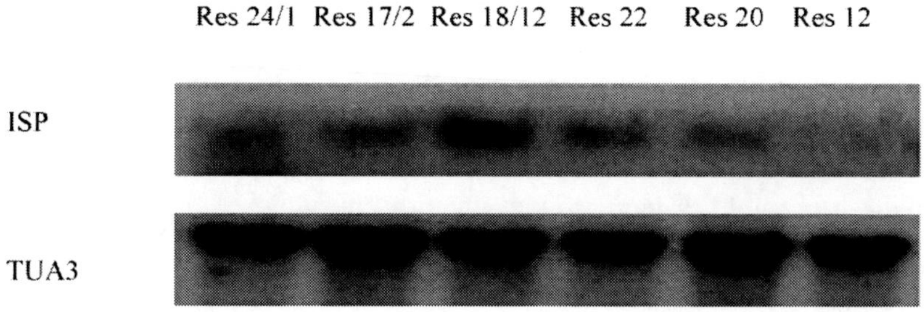

Figure 1. Results of Nothern blot hybridization of transgenic potato lines (Рез-24/1, Рез-17/2 and Рез-18/12) and tobacco (C20, C22 and C12). *ISP* – gene of serine proteinase inhibitor from buckwheat seeds, *TUA3* – gene of α-tubulin from arabidopsis, used as standard.

Further Nothern blot analysis has been carried out for detection of mRNA transcript of the gene of proteinase inhibitor in obtained transgenic tobacco and potato lines as well as for determination of the level of gene expression. Fragment of vector construction of $1.2.10^3$ bp, containing the sequence of *ISP* gene, was used as the sound. DNA-sound of 1.8. 10^3 bp, containing α-tubulin TUA3 gene from arabidopsis was used as control in the course of Nothern hybridization. mRNA transcript of expected value (213 bp) was detected in transgenic potato lines Pe3-24/1, Pe3-17/2, Pe3-18/12 and potato C22 and C20. Fig. 1 demonstrates that expression of *ISP* gene was different for various potato and tobacco lines. mRNA amount was the highest in potato line Pe3-18/12, whereas potato lines Pe3-17/2, Pe3-24/1 and tobacco line C20 showed a lesser expression level of *ISP* gene. Almost no expression of ISP gene was found in tobacco C12 line, which lost tolerance to kanamycin. This may point either to elimination of the *ISP* gene or to gene silence.

Analysis of obtained transformed plants indicated that transformation frequency was increased when mixed cultivation and regeneration were performed on the cysteine containing medium. It is noteworthy that the used cysteine concentration of 400 μg/ml inhibited growth of agrobacteria what increased the duration of subcultivation and decreased their number on media containing antibiotics. Possibly, it stimulated also formation of more regenerated plants.

Since all primary transformants formed roots on medium with kanamycin their sensitivity to kanamycin in the course of callus formation and morphogenesis (de-differentiation and re-differentiation) was additionally studied. Explants of all studied lines formed callus and roots in the media for callus formation and preserved ability to form adventive sprouts on regeneration media with kanamycin. Besides, the lines retained ability to grow and regenerate on media with kanamycin even after multiple subcultivation and regeneration on media without kanamycin. At the same time processes of callus formation and morphogenesis for control plant were observed only on media devoid of selective agent.

Individual tobacco and potato lines, transformed by vector construction with inhibitor gene, were analyzed by PCR technique. 90.5% of tobacco regenerates and 60% of potato regenerates gave positive result after PCR analysis. By means of the usual technique by Horsh et al. (Horsch et al. 1987) the transformation frequency of tobacco and potato leaf segments was established (43% and 53%, respectively).

Thus, use of technique of mixed cultivation and regeneration in the presence of cysteine significantly increased frequency as well as efficiency of transformation of both plants. The large amount of transgenic lines obtained in the course of transformation of tobacco can be explained by the chosen selection scheme with a high content of kanamycin (150 μg/ml), what strongly decreased the number of regenerated pseudotransformed plants. It turned out that it was impossible to perform selection of potato plants at so high kanamycin concentrations in view of complete inhibition of plant growth. Thus, the use of the mentioned technique made it possible to increase significantly the yield of regenerates in the whole as well as the number of transformed plants, apparently, because of reduction of necrotizing tissues and increase in stable transfer of T-DNA into plant cells.

Antibacterial activity of transgenic lines was determined using dish tests. Tissues of non-transformed control tobacco plant ($Control_{nt}$) and that transformed by the *ISP* gene-free vector ($Control_0$) practically did not inhibit growth of any used bacterium, whereas tissues of all analyzed transgenic plants exhibited antibacterial activity (Table 2).

Evidently, transformants synthesized a target functional protein exhibiting inhibitory effect on bacterial growth. However, significant differences were detected in the extent of inhibition of different bacterial species. It should be noted that selection of bacterial strains *C.*

michiganensis and *P. syringae* is explained by the fact that they secrete serine proteinases during their vital activities (Burger et al. 2005; Jang et al. 1996). Transgenic clones inhibited growth of *P. syringae* and to somewhat lower extent growth of *C. michiganensis*. Besides. secondary growth of separate *C. michiganensis* colonies was always observed in the inhibition zone. To the least extent transgenic clones inhibited growth of *E. coli*; in this case even not all transgenic lines exhibited inhibitory effect. It should be noted that tissues of C12 line, that lost their resistance to kanamycin after passage for 2 years, also lost the ability to inhibit bacterial growth (Table 2).

Table 2. Antibacterial activity of tissues of primary tobacco transformants.

Line	Mean diameter of the zone of bacteria growth inhibition including diameter of the well, cm		
	E. coli	*Pseudomonas syringae*	*Clavibacter michiganensis*
$Control_{nt}$	0.5*	0.5*	0.5*
$Control_{O}$	0.5*	0.5*	0.5*
C1	1.10±0.08	2.27±0.24	1.28±0.08
C2	0.71±0.04	1.63±0.13	1.23±0.11
C7	0.62±0.02	1.82±0.21	1.20±0.07
C8	0.80±0.03	2.09±0.18	1.25±0.13
C10	1.30±0.24	1.92±0.17	1.26±0.11
C11	0.85±0.11	1.98±0.19	1.24±0.09
C12	0.5*	0.5*	0.5*
C15	1.00±0.15	1.81±0.22	1.32±0.13
C16	1.35±0.22	1.84±0.12	1.21±0.07
C18	1.02±0.41	1.77±0.16	1.56±0.21
C19	0.71±0.11	1.78±0.26	1.18±0.08
C20	1.11±0.27	1.92±0.15	1.43±0.16
C22	0.63±0.08	2.18±0.19	1.63±0.11
C37	1.00±0.09	1.80±0.18	1.15±0.08
C41	1.21±0.14	1.76±0.14	1.31±0.14
C44	0.92±0.10	1.92±0.23	1.52±0.13
C47	1.31±0.15	2.01±0.19	1.42±0.09

*Diameter of the well - 0.5 cm.

Tissues of transgenic lines of different species highly differed in the extent of antibacterial activity. Thus. wells containing tissue homogenates of transgenic tobacco plants on the lawn of *E. coli* model culture were surrounded by a halo of growth inhibition of 1.5-1.7 cm in diameter (including the well diameter). In potato, protective effect was much more pronounced in the case of application of the same amount of homogenized tissue, and the halo diameter reached 3-5 cm . Probably, so pronounced difference in the growth inhibition zone dimensions is caused by individual biological peculiarities of each plant, and possibly by different copy numbers of vector constructs inserted into their genomes.

Exposure of tobacco plants by the glasshouse whiteflies (*Trialeurodes vaporariorum*) observed in the green house has shown that transgenic plants were resistant to the effect of these insects, whereas non-transformed ones were strongly affected by them. Numerous eggs were laid by the whitefly on the control plant leaves and full-value insects emerged from these eggs (Fig. 2B). Only single ovipositions of 1-2 eggs were observed on transgenic plants, and no progeny appeared in this case (Fig. 2A).

It should be noted that unlike a number of different insects, serine proteinases are prevalent in the whitefly alimentary canal (Terra and Ferreira 1994). Thus, introduction of just a single gene of serine proteinase inhibitor into the plants of heterologous group has shown the possibility of obtaining protective effects against insects and phytopathogenic bacteria.

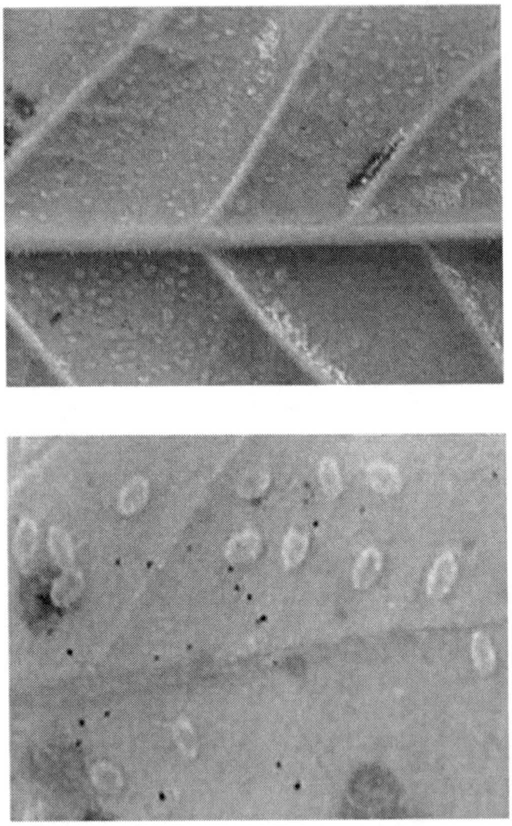

A

B

Figure 2. Affection by the whitefly of the control tobacco plant (B), and the absence of egg laying on transgenic plant (A) in the greenhouse.

The results indicate that, unlike control plants, those grown by us are able to synthesize certain functional proteins that exhibit protective effect and inhibit development of bacteria. It appeared that the presence of just a single gene of serine proteinase inhibitor provides for sufficient protection against at least two phytopathogenic bacteria. It is quite important to note that to a higher extent transgenic clones inhibited growth of phytopathogenic bacteria *P. syringae* and *C. michiganensis* and to a significantly lower extent growth of model culture *E. coli*. which is probably due to the specificity of proteolytic enzymes of different bacterial species. In addition. significant genotypic distinctions in protection efficiency were observed between members of different genera of the same family (potato and tobacco), which requires further investigation for final clarification of this question. Experiments on potato transgenic lines carrying the *ISP* gene of buckwheat proteinase inhibitor revealed fungicidal effect on *Fusarium sp*., resulted in cessation of fungus growth at a distance >5 mm from the transgenic line explants, whereas control (non-transformed) variants were completely overgrown (Tcherednitchenko 2004).

Note that the purified preparation of this protein-inhibitor at concentrations of 0.05-0.5 μg/ml inhibited by 50-100% growth of hyphae of the filamentous fungus *A. alternata* (Dunaevsky et al. 1998).

CONCLUSION

Thus, the possibility of involvement of the recombinant plant inhibitor of proteolytic enzymes in protection of various plants against pathogenic microorganisms and insects is shown in this work. In connection with recent achievements of biotechnology in creation of genetically modified plants characterized by increased resistance to different unfavorable effects, such approach becomes more and more actual because it not only enables the increase in productivity of cultured plants but contributes to improvement of ecological conditions due to reduced usage of highly toxic protective chemicals.

To better understand the mechanism of interactions between a phytopathogen and the host plant, it is necessary to intensify the search for targets of inhibitors in mycelium fungi. Progress in this field, along with the availabe data, will allow efficient engineering of transgenic plants with a higher resistance to pathogenic microflora on the basis of protease inhibitor genes.

The observed differences in the extent of growth inhibition of different bacterial species are also of undoubted interest for further investigation of possible mechanisms of antibacterial activity of transgenic lines expressing this inhibitor.

ACKNOWLEDGMENTS

This work was supported by the Russian Foundation for Basic Research (grants 09-04-00955, 10-04-00739) and International Scientific and Technology Center (grant 3455).

REFERENCES

[1] Ahn, J.-E., Salzman, R.A., Braunagel, S.C., Koiwa, H. & Zhu-Salzman, K. (2004). Functional roles of specific bruchid protease isoforms in adaptation to a soybean protease inhibitor. *Insect Mol. Biol.*, 13, 649-657.

[2] Alfonso-Rubi, J., Ortego, E., Castanera, P., Carbonero, P. & Diaz, I. (2003). Transgenic expression of trypsin inhibitor CMe from barley in *indica* and *japonica* rice, confers resistance to the rice weevil *Sitophilus oryzae. Transgenic Res.,* 12, 23-31.

[3] Altpeter, F., Diaz, I. McAuslane, H., Gaddour, K., Carbonero, P. & Vasil, I.K. (1999). Increased insect resistance in transgenic wheat stably expressing trypsin inhibitor CMe. *Mol. Breed.*, 5, 53-63.

[4] Benken, I.I.; Mosolov, V.V. & Fedurkina, N.V. (1976). Effects of an inhibitor of bean proteinase on plant pathogenic fungi. *Mycology and Phytopathology*, 10, 198-201

[5] Bouchard, E., Cloutier, C. & Michaud, D. (2003). Oryzacystatin 1 expressed in transgenic potato induces digestive compensation in an insect natural predator via its herbivorous prey feeding on the plant. *Mol. Ecol.*, 12, 2439-2446.

[6] Boulter, D., Edwards, G.A., Gatehouse, A.M.R., Gatehouse, J.A. & Hilder, V.A. (1990). Additive protective effects of incorporating two different higher plant derived insect resistance genes in transgenic tobacco plants. *Crop Prot.*, 9, 351-354.

[7] Brunelle, F., Nguyen-Quoc, B., Cloutier, C. & Michaud, D. (1999). Protein hydrolysis by colorado potato beetle, *Leptinotarsa decemlineata*, digestive proteases: the catalytic role of cathepsin D. Arch. *Insect Biochem. Physiol.*, 42, 88-98.

[8] Bu, Q-Y, Wu, L, Yang, S-H. & Wan, J-M. (2006). Cloning of a potato proteinase inhibitor gene PINII-2x from diploid potato (*Solanum phurejia* L.) and transgenic investigation of its potential to confer insect resistance in rice. *J. Integrative Plant Biol.*, 48, 732-739

[9] Burger, A., Graffen, I., Engelman, J., Niermann, E., Pieper, M., Kircher, O., Gertemann, K.H. & Eichenlaub, R. (2005). Identification of homologues to the pathogenicity factor Pat-1, a putative serine protease of *Clavibacter michiganensis* subsp. *michiganensis*. *Microbiol. Res.,* 160, 417–427.

[10] Charity, J.A., Anderson, M.A., Bittisnich, D.J., Whitecross, M. & Higgins, T.J.V. (1999). Transgenic tobacco and peas expressing a proteinase inhibitor from *Nicotiana alata* have increased insect resistance. *Mol. Breed.*, 5, 357-365.

[11] Charity, J.A., Hughes, P., Anderson, M.A., Bittisnich, D.J., Whitecross, M. & Higgins, T.J.V. (2005). Pest and disease protection conferred by expression of barley b-hordothionin and *Nicotiana alata* proteinase inhibitor genes in transgenic tobacco. *Func. Plant Biol.,* 32, 35–44.

[12] Chen, Z.Y., Brown, R.L., Lax, A.R., Cleveland, T.E. & Russin, J.S. (1999). Inhibition of plant pathogenic fungi by a corn trypsin inhibitor over-expressed in *Escherichia coli*. *Appl. Environ. Microbiol.*, 65, 1320–1324.

[13] Cloutier, C., Jean, C., Foumier, M., Yelle, S. & Michaud, D. (2000). Adult Colorado potato beetles, *Leptinotarsa decemlineata,* compensate for nutritional stress on oryzacystatin 1-transgenic potato plants by hypertrophic behavior and over-production of insensitive proteases. *Arch. Insect Biochem. Physiol.,* 44, 69-81.

[14] Cordero, M.J., Raventos, D. & San Segundo, B. (1994). Expression of a maize proteinase inhibitor gene is induced in response to wounding and fungal infection: systemic wound-response of a monocot gene. *Plant J.,* 6, 141–150.

[15] Delledonne, M., Allegro, G., Belenghi, B., Balestrazzi, A., Picco, F., Levine, A., Zelasco, S., Calligari, P. & Confalonieri, M. (2001). Transformation of white poplar (*Populus alba* L.) with a novel Arabidopsis thaliana cysteine proteinase inhibitor and analysis of insect pest resistance. *Mol. Breed.*, **7,** 35-42.

[16] De Leo, F., Bonade-Bottino, M., Ceci, L.R., Gallerani, R. & Jouanin L. (1998). Opposite effects on *Spodoptera littoralis* larvae of high expression level of a trypsin proteinase inhibitor in transgenic plants. *Plant Physiol.*, 118, 997-1004.

[17] De Leo, F., Bonadè-Bottino, M., Ceci, L.R., Gallerani, R. & Jouanin, L. (2001). Effects of a mustard trypsin inhibitor expressed in different plants on three lepidopteran pests. Insect Biochem. *Mol. Biol.*, 31, 593-602.

[18] De Leo, F. & Gallerani, R. (2002). The mustard trypsin inhibitor 2 affects the fertility of *Spodoptera littoralis* larvae fed on transgenic plants. Insect Biochem. *Mol. Biol.* 32, 489-496.

[19] Dunaevsky, Y.E., Pavlukova, E.B., Beliakova, G.A., Tsybina, T.A., Gruban, T.N. & Belozersky, M.A. (1998). Protease inhibitors in buckwheat seeds: comparison of anionic and cationic inhibitors. *J. Plant Physiol.*, 152, 696-702.

[20] Edmonds, H.S., Gatehouse, L.N., Hilder, V.A. & Gatehouse, J.A. (1996). The inhibitory effects of the cysteine protease inhibitor , oryzacystatin, on digestive proteases and on larval survival and development of the southern corn rootworm (*Diabrotica undecim punctata* howard). *Entomol. Exp. Appl.*, 78, 83-94.

[21] Falco, M.C. & Silva-Filho, M.C. (2003). Expression of soybean proteinase inhibitors in transgenic sugarcane plants: effects on natural defense against *Diatraea saccharalis. Plant Physiol. Biochem.*, 41, 761-766.

[22] Gatehouse, A.M.R., Hilder, V.A., Powell, K.S., Boulter, D. & Gatehouse, J.A. (1992). *Potential of plant-derived genes in the genetic manipulation of crops for insect resistance*. In: Proc. 8th Int. Symp. Insect-Plant Relationships. Dordrecht: Kluwer Acad. Publ. p. 221-234

[23] Gatehouse, A.M.R., Shi, Y., Powell, K.S., Brough, C., Hilder, V.A., Hamilton, W.D.O., Newell, C.A., Merryweather, A., Boulter, D. & Gatehouse, J.A. (1993). Approaches to insect resistance using transgenic plants. *Phil. Trans. R. Soc. Lond.*, 342, 279-286.

[24] Girard, C., Le Metayer, M., Bonade-Bottino, M., Pham-Delegue, M.H. & Jouanin, L. (1998). High level of resistance to proteinase inhibitors may be conferred by proteolytic cleavage in beetle larvae. *Insect Biochem. Mol. Biol.*, 28, 229-237.

[25] Gorbalenya, A.E. & Snijder, E.J. (1996). Viral cysteine proteinases. *Perspect. Drug. Discov. Design.*, 6. 64–86.

[26] Hilder, V.A., Gatehouse, A.M.R., Sheerman, S.E., Barker, R.F. & Boulter, D. (1987). A novel mechanism of insect resisrance engineered into tobacco. *Nature*, 330, 160-163.

[27] Hoffmann, M.P., Zalom, F.G., Wilson, L.T., Smilanick, J.M., Malyj, L.D., Kiser, J., Hilder, V.A. & Barnes, W.M. (1992). Field evaluation of transgenic tobacco containing genes encoding *Bacillus thuringiensis* δ-endotoxin or cowpea trypsin inhibitor: efficacy against *Helicoverpa zea* (Lepidoptera: Noctuidae). *J. Econ. Entomol.*, 85, 2516-2522.

[28] Horsch, R.B., Fry, J.T., Hoffman, N.L., Einholtz, D., Rogers, S.G. & Fraley, R.T. (1987). A simple and general method for transferring genes into plants. *Science*, 227, 1229–1230.

[29] Jang, W.H., Kim, E.K., Lee, H.B., Chung, J.H. & De, O.J. (1996). Characterization of an alkaline serine protease from an alkaline-resistant *Pseudomonas* sp.: cloning and expression of the protease gene in *Escherichia coli*. *Biotechnol. Lett.*, 18, 57–62.

[30] Joshi, B.N., Sainani, M.N., Bastawade, K.B., Gupta, V.S. & Ranjekar, P.K. (1998). Cysteine protease inhibitor from pearl millet: a new class of antifungal protein. *Biochem. Biophys. Res. Commun.*, 246, 382-387.

[31] Kladnistskaya G.V., Valueva T.A., Ermolova N.V., Ilyinskaya L.I., Gerasimova N.G., Mosolov V.V. (1996). Accumulation of proteinase inhibitors in potato tuber diffusates under infection with phytophtorosis pathogen. Rus. *J. Plant Physiol.* 43, 701-706.

[32] Koiwa, H., Shade, R.E., Zhu-Salzman, K., D'Urzo, M.P., Murdock, L.L., Bressan, R.A. & Hasegawa, P.M. (2000). A plant defensive cystatin (soyacystatin) targets cathepsin 1-like digestive cysteine proteinases (DvCALs) in the larval midgut of western corn rootworm (*Diabrotica virgifera virgifera). FEBS Lett*., 471, 67-70.

[33] Konarev, A.V. (1987). Variation of trypsin-like proteinase inhibitors in wheat and related grain crops in connection with grain pest resistance. *Agr. Biol.*, 5, 17-24.

[34] Kondo, H., Ijiri, S., Abe, K., Maeda, H. & Arai, S. (1992). Inhibitory effect of oryzacystatins and a truncation mutant on the replication of poliovirus in infected Vero cells. *FEBS Lett.*, 299, 48–50.

[35] Lawrence, P.K. & Koundal, K.R. (2002). Plant protease inhibitors in control of phytophagous insects. Electron. J. Biotechnol., 5, 93-109.

[36] Lecardonnell, A., Chauvin, L., Jouanin, L. Beaujean, A., Prevost, G. & Sangwan-Norreel, B. (1999). Effects of the rice cystatin 1 expression in transgenic potato on Colorado potato beetle larvae. *Plant Sci.*, 140, 71-79.

[37] Lee, S.I., Lee, S.H., Koo, J.C., Chun, H.J., Lim, C.O., Mun, J.H., Song, Y.H. & Cho, M.J. (1999). Soybean Kunitz trypsin inhibitor (SHTI) confers resistance to the brown planthopper (*Nilaparvata lugens* Stal) in transgenic rice. *Mol. Breed.* 5, 1-9.

[38] Lorito, M., Broadway, R.M., Hayes, C.K., Woo, S.L., Noviello, C., Williams, D.L. & Harman, G.E. (1994). Proteinase inhibitors from plants as a novel class of fungicides. *Mol. Plant-Microbe Interact*., **7**, 525-527.

[39] Luo, M., Wang, Z., Li, H., Xia, K-F., Cai, Y. & Xu, Z-F. (2009). Overexpression of a Weed (*Solanum americanum*) Proteinase Inhibitor in Transgenic Tobacco Results in Increased Glandular Trichome Density and Enhanced Resistance to *Helicoverpa armigera* and *Spodoptera litura*. *Int. J. Mol. Sci*., 10, 1896–1910.

[40] MacIntosh, S.C., Kishore G.M., Perlak F.J., Marrone P.G., Stone T.B., Sims S.R., Fuchs R.L. 1990. Potentiation of *Bacillus thuringiensis* insecticidal activity by serine protease inhibitors. *J. Agric. Food Chem.,* 38, 50-58.

[41] MacManus, M.T. & Burgess, E.P.J. (1999). Expression of the soybean (Kunitz) trypsin inhibitor in transgenic tobacco: Effects on larval development of *Spodoptera litura. Transgenic Res.*, **8**, 383-395.

[42] Marchetti, S., Delledonne, M., Fogher, C., Chiaba, C., Chiesa, F., Savazzini, F. & Glordano, A. (2000). Soybean Kunitz, C-II and PI-IV inhibitor genes confer different levels of insect resistance to tobacco and potato transgenic plants. *Theor. Appl. Genet*., 101, 519-526.

[43] Mazumdar-Leighton, S., & Broadway, R.M. (2001a). Identification of six chymotrypsin cDNAs from larval midguts of *Helicoverpa zea* and *Agrotis ipsilon* feeding on the soybean (Kunitz) trypsin inhibitor. *Insect Biochem. Mol. Biol.,* 31, 633-644.

[44] Mazumdar-Leighton, S. & Broadway, R.M. (2001b). Transcriptional induction of diverse midgut trypsins in larval *Agrotis ipsilon* and *Helicoverpa zea* feeding on the soybean trypsin inhibitor. *Insect Biochem. Mol. Biol.*, 31, 645-657.

[45] Michaud, D., Nguyen-Quoc, B., Vrain, T.C., Fong, D. & Yelle S. (1996). Response of digestive proteinases from the Colorado potato beetle (*Leptinotarsa decemlineata*) and the black vine weevil (*Otiorynchus sulcatus*) to a recombinant form of human stefin A. *Arch. Insect Biochem. Physiol.*, 31, 451-464.

[46] Mosolov, V.V. & Valueva, T.A. (1993). *Plant Protein Inhibitors of Proteolytic Enzymes*. (Ed. Kretovich, V.L.). Moscow: VINITI.

[47] Ninković, S., Miljuš-Đukić, J., Radović, S., Maksimović, V., Lazarević, J., Vinterhalter, B., Nešković, M. & Smigocki, A. (2007). *Phytodecta fornicata* Brüggemann resistance mediated by oryzacystatin II proteinase inhibitor transgene. *Plant Cell, Tissue and Organ Culture,* 91, 289-294.

[48] Olhoft, P.M. & Somers, D.A. (2001). L-Cysteine increases *Agrobacterium*-mediated T-DNA delivery into soybean cotyledonary-node cells. *Plant Cell Rep.*, 20, 706-711

[49] Orozco-Cardenas, M., McGurl, B. & Ryan, C. (1993). Expression of an antisense prosystemin gene in tomato plants reduces resistance toward *Manduca sexta* larvae. *Proc. Natl. Acad. Sci. USA*, 90, 8273-8276.

[50] Ortego, F., Sanchez-Ramos, I., Ruiz, M. & Castanera, P. (2000). Characterization of proteases from a stored product mite *Tyrophagus putrescentiae. Arch. Insect Biochem. Physiol.,* 43, 116-124.

[51] Outchkourov, N.S., de Kogel, W.J., Schuurman-de Bruin, A., Abrahamson, M. & Jongsma, M.A. (2004a). Specific cysteine protease inhibitors act as deterrents of western flower thrips, *Frankliniella occidentalis* (Pergande), in transgenic potato. *Plant Biotechnol. J.*, 2, 439-448.

[52] Outchkourov, N.S., de Kogel, W.J., Wiegers, G.L., Abrahamson, M. & Jongsma, M.A. (2004b). Engineered multidomain cysteine protease inhibitors yield resistance against western flower thrips *(Frankliniella occidentalis)* in greenhouse trials. *Plant Biotechnol. J.*, 2, 449-458.

[53] Peng, J.H. & Black, L.L. (1976). Increased proteinase inhibitor activity in response to infection of resistant tomato plants by *Phytophthora infestans*. *Phytopathol.*, 66, 958–963.

[54] Pernas, M., Sanchez-Monge, R. & Salcedo, G. (2000). Biotic and abiotic stress can induce cystatin expression in chestnut. *FEBS Lett.*, 467, 206–210.

[55] Roby, D., Toppan, A. & Esquerre-Tugaye, M.T. (1987). Cell surfaces in plant microorganisms interactions. VIII. Increased proteinase inhibitor activity in melon plants in response to infection by Colletotrichum lagenarium or to treatment with an elicitor fraction from this fungus. *Physiol. Mol. Plant Pathol.,* 30, 453–460.

[56] Ryan, C.A. (1990). Protease inhibitors in plants: genes for improving defenses against insects and patogens. *Annu. Rev. Phytopathol.*, 28, 425-449.

[57] Srinivasan, T., Kumar, K.R. & Kirti, P.B. (2009). Constitutive expression of a trypsin protease inhibitor confers multiple stress tolerance in transgenic tobacco. *Plant Cell Physiol.,* 50, 541-553.

[58] Tcherednitchenko, M.Y. (2004). *Transformation of potato and tobacco by genes of defensins and proteinase inhibitor BWI-1a.* PhD thesis, Russian State Agricultural University, Moscow, 134 pp.

[59] Terra, W.R., & Ferreira, C. (1994). Insect digestive enzymes: properties, compartmentalization and function. *Comp. Biochem. Physiol.*, B 109, 1–62.

[60] Thomas, M.B. (1999). *Ecological approaches and the development of "truly integrated" pest management.* Proc. Natl. Acad. Sci. USA, 96, 5944-5951.

[61] Thomas, J.C., Adams, D.G., Keppenne, V.D., Wasmann, C.C., Brown, J.K., Kanost, M.R. & Bohnert, H.J. (1995). Protease inhibitors of *Manduca sexta* expressed in transgenic cotton. *Plant Cell Rep.*, 14, 758-762.

[62] Valueva T.A., Kladnitskaya G.V., Ilyinskaya L.I., Gerasimova N.G., Ozeretskovskaya O.L., Mosolov V.V. (1998) Chymotrypsin inhibitorsn in potato tubers infected with phytophtorosis pathogen. *Rus. J. Bioorg. Chem*. 24, 346-349.

[63] Van Loon, L.C. (1985). Pathogenesis-related proteins. *Plant Mol. Biol.,* 4, 111–116.

[64] Vila, L., Quilis, J., Meynard, D., Breitler, J.C., Marfa, V., Murillo, I., Vassal, J.M., Messeguer, J., Guiderdoni, E. & San Segundo, B. (2005). Expression of the maize proteinase inhibitor (mpi) gene in rice plants enhances resistance against the striped stem borer (*Chilo suppressalis*) : effects on larval growth and insect gut proteinases. *Plant Biotechnol. J.,* 3, 187-202.

[65] Wu, Y.R., Liewellyn, D., Mathews, A. & Dennis, E.S. (1997).. Adaptation of *Helicoverpa armigera* (Lepidoptera: Noctuidae) to proteinase inhibitor expressed in transgenic tobacco. *Mol. Breed*., 3, 371-380.

[66] Xu, Z.F., Teng, W.L. & Chye, M.L. (2004). Inhibition of endogenous trypsin- and chymotrypsin-like activities in transgenic lettuce expressing heterogeneous proteinase inhibitor SaPIN2a. *Planta*, 218**,** 623-629.

[67] Zhu-Salzman, K., Koiwa, H., Salzman, R.A., Shade, R.E. & Ahn, J.-E. (2003). Cowpea bruchid *Callosobruchus maculatus* uses a three-component strategy to overcome a plant defensive cysteine protease inhibitor. *Insect Mol. Biol.***,** 12, 135-145.

INDEX

C

D

E

F

N

O

P

Q

R

V

W

Y

Z